North Yorkshire County Council Library Service
Renew online at www.northyorks.gov.uk/libraries

THIS IS A WELBECK BOOK

First published in 2020 by Welbeck,
an imprint of Welbeck Non-Fiction Limited,
part of the Welbeck Publishing Group
20 Mortimer Street
London W1T 3JW

Design © Welbeck Non-Fiction Limited 2020
Text copyright © Welbeck Non-Fiction Limited 2020

All rights reserved. This book is sold subject to the condition that it may not be reproduced, stored in a retrieval system or transmitted in any form or by any means, electronic, mechanical, photocopying, recording or otherwise without the publisher's prior consent.

A CIP catalogue for this book is available from the British Library.

ISBN 978-1-78739-531-2

Printed in Dubai

10 9 8 7 6 5 4 3 2 1

CONTAGION

PLAGUES, PANDEMICS AND CURES FROM THE BLACK DEATH TO COVID-19 AND BEYOND

DR RICHARD GUNDERMAN

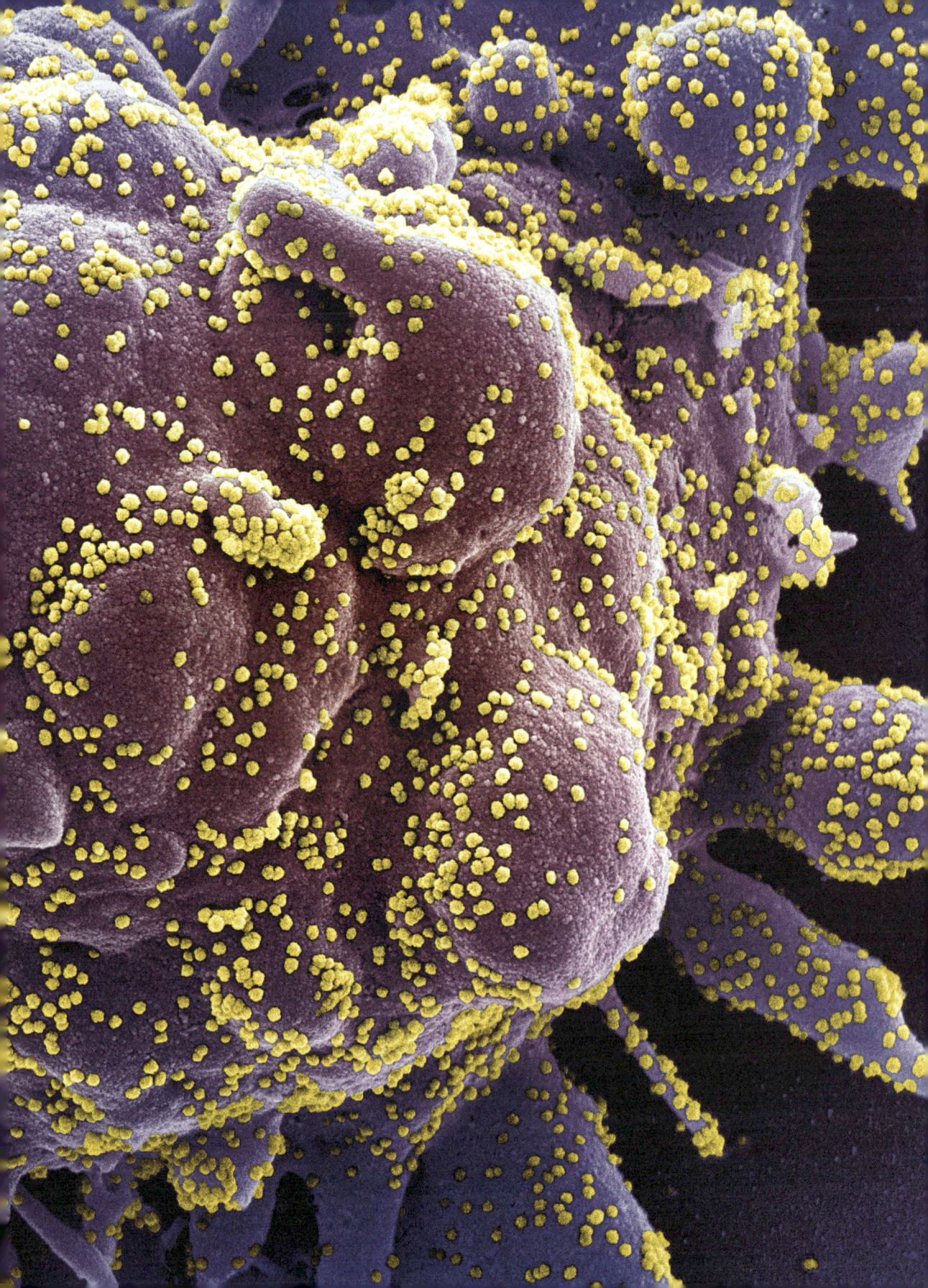

CONTENTS

1. Reports of the Death of Infectious Disease were Greatly Exaggerated — 6
2. Infectious Disease — 10
3. The Life of an Infectious Microbe — 14
4. Natural Selection and Infectious Disease — 18
5. Ancient Views of Health and Disease: The Hippocratics — 24
6. The Plague of Athens — 28
7. The Black Death — 32
8. Boccaccio and the Black Death — 38
9. Spain's Conquest of the Aztecs — 44
10. Seeing Microbes for the First Time: Leeuwenhoek — 50
11. Smallpox Inoculation and the American Founding — 54
12. The Physician as Hero in Plague Time: Dr Benjamin Rush — 58
13. Edward Jenner and the Little Prick — 64
14. Tuberculosis, the Persistent Killer — 70
15. Tuberculosis: A Poetic Case — 76
16. John Snow, Founder of Epidemiology — 80
17. Ignaz Semmelweis, Apostle of Handwashing — 84
18. Joseph Lister, Microbe Killer — 86
19. Florence Nightingale, the "Lady With The Lamp" — 90
20. Pasteur, Microbiologist Extraordinaire — 94
21. Robert Koch and his Radical Postulates — 98
22. Pettenkofer, Good Effects from Wrong Ideas — 102
23. The "Greatest Pandemic In History" — 104
24. The World's Deadliest Animal — 112
25. STIs: Sexually Transmitted Infections — 118
26. Penicillin — 122
27. Attempts to Eradicate Infectious Disease — 126
28. HIV/AIDS — 130
29. Is Peptic Ulcer Disease Infectious? — 136
30. Vaccinating Against Cancer: HPV — 140
31. Infectious Disease as a Weapon: Bioterrorism — 142
32. Coronaviruses: Twenty-First-Century Pandemic Kings — 146
33. Infectious Disease: The Road Forward — 152

Further Reading — 156
Index — 157
Credits — 160

REPORTS OF THE DEATH OF INFECTIOUS DISEASE WERE GREATLY EXAGGERATED

BATTLES WON

It is undeniable that the world has made great progress in reducing the toll of many once-prominent infectious diseases.

Smallpox, which once ranked as one of the major causes of infant mortality and, at the time of the founding of the United States, accounted for 10 per cent of all deaths, is no longer being transmitted from person to person. Building on one of the most important innovations in the history of the battle against infectious disease, immunization, public health workers have been able to eradicate the disease from the face of the earth, and the virus is now thought to exist only in laboratories.

During the nineteenth century, many major cities were wracked by epidemics of cholera, producing calls for improved water and sanitation systems and widespread use of quarantine. By no means has the disease been completely eradicated, but at least in wealthier societies, cases of the disease are rare. The epidemiological studies of John Snow and microbiological work of Robert Koch elucidated the causative organism and its modes of transmission, permitting effective control measures.

Steps taken against one disease often help to reduce the burden of other diseases. The improvements in water and sanitation systems undertaken to lessen the incidence of cholera paid additional dividends in the reduction of typhoid fever. The existence of asymptomatic carriers such as the notorious "Typhoid Mary" can now usually be addressed by high doses of antibiotics and the removal of the gallbladder, where the causative microbes often reside.

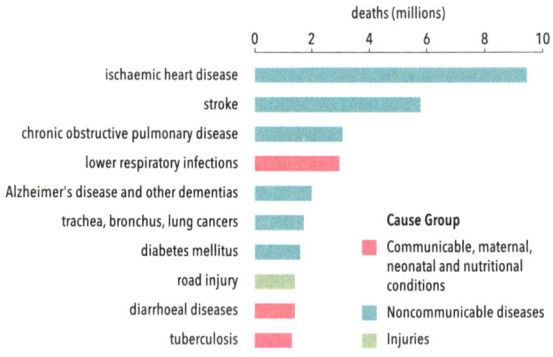

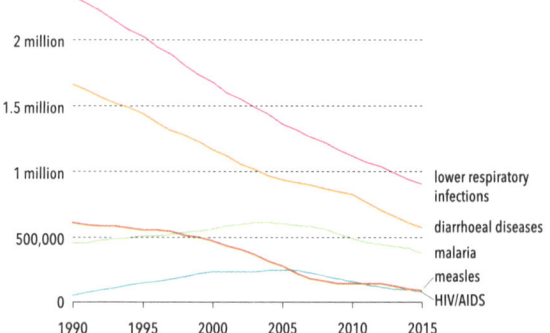

Malaria, a disease that causes intermittent fevers, was long blamed on noxious vapours or miasma that arose from swamps. When Ronald Ross discovered that mosquitoes transmit the disease, securing him the 1902 Nobel Prize in Physiology or Medicine, the strategy for eradicating the disease became clear: draining

Reports of the Death of Infectious Disease Were Greatly Exaggerated

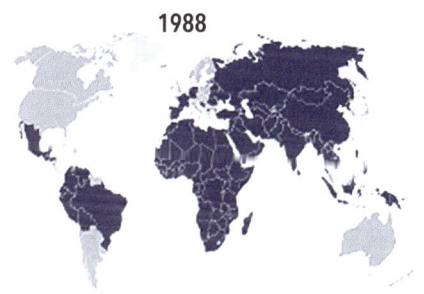

collections of fresh water where mosquitoes breed. Later, insecticides also made an important contribution. Today, many wealthy nations report no cases of the disease.

Yellow fever is another disease that is transmitted by mosquitoes, as proved by Walter Reed, who divided volunteers into groups that allowed themselves to be bitten by mosquitoes and others who were protected from them. Only those who were bitten developed the disease. Again, because mosquitoes were the common vector, synergies resulted from efforts to reduce malaria and yellow fever. Max Theiler received the 1951 Nobel Prize for developing the first attenuated virus vaccine.

Poliomyelitis once ravaged temperate climates, causing death and permanent paralysis in many of its victims. Interest in the disease was stimulated by its most famous patient, US President Franklin Roosevelt. Eventually, the development of vaccines, financed in the US by the March of Dimes campaign, enabled Jonas Salk to develop an inactivated vaccine, followed by Alfred Sabin's attenuated vaccine, leading to the near-eradication of the disease in wealthy countries.

Tuberculosis reigned as the major cause of death in the Western world for centuries, and it was responsible for up to a quarter of deaths in large urban centres. The isolation of the tubercle bacillus by Robert Koch secured him the 1905 Nobel Prize. Routes of transmission such as infected milk were eliminated, and general improvements in diet and housing helped to drive down the infection rate, followed by the development of effective drugs. Today the disease is rare in many wealthy nations.

Above, from left: Two communicable diseases: a cartoon from 1886 showing the dangers of polluted water and cholera, and a smallpox quarantine poster from 1910s San Francisco.

■ countries that have never eliminated polio
■ countries that have eliminated polio

AN ONGOING WAR

Despite these and other successes, infectious diseases have by no means been eradicated. Death rates due to infection remain high in many poorer parts of the world, where diarrhoeal diseases and malaria remain major causes of death, and diseases of centuries past such as measles, pertussis, and tetanus continue to take a major toll. Many of these diseases, including the organisms that cause pneumonia, hit children particularly hard.

Moreover, new diseases have arisen. HIV/AIDS, which was first identified in the 1980s, is thought to represent the world's number-two infectious killer, resulting in several million deaths per year. Among the problems this disease presents is the fact that many infected patients remain largely asymptomatic for years after infection, during which they can transmit it to others. Untreated individuals die on average about ten years after infection.

The number-one cause of death from infectious disease is lower respiratory infections, especially pneumonia and bronchitis. But beneath this apparently simply fact lies a large group of pathogens, including streptococcal and staphylococcal bacteria, parasites such as cryptosporidiosis, and a large variety of viral pathogens, including influenza A and B, adenovirus, parainfluenza virus, and respiratory syncytial virus.

Particularly vexing are viral causes of pneumonia, such as influenza viruses, which mutate regularly. As a result, a strain of flu that spreads across the world one year may provide no immunity to a subsequent strain the next year. Hence people are encouraged to receive annual flu immunizations against the most likely strains. Despite its ubiquity and regularity, seemingly simple aspects of influenza remain poorly understood, such as why outbreaks tend to be seasonal.

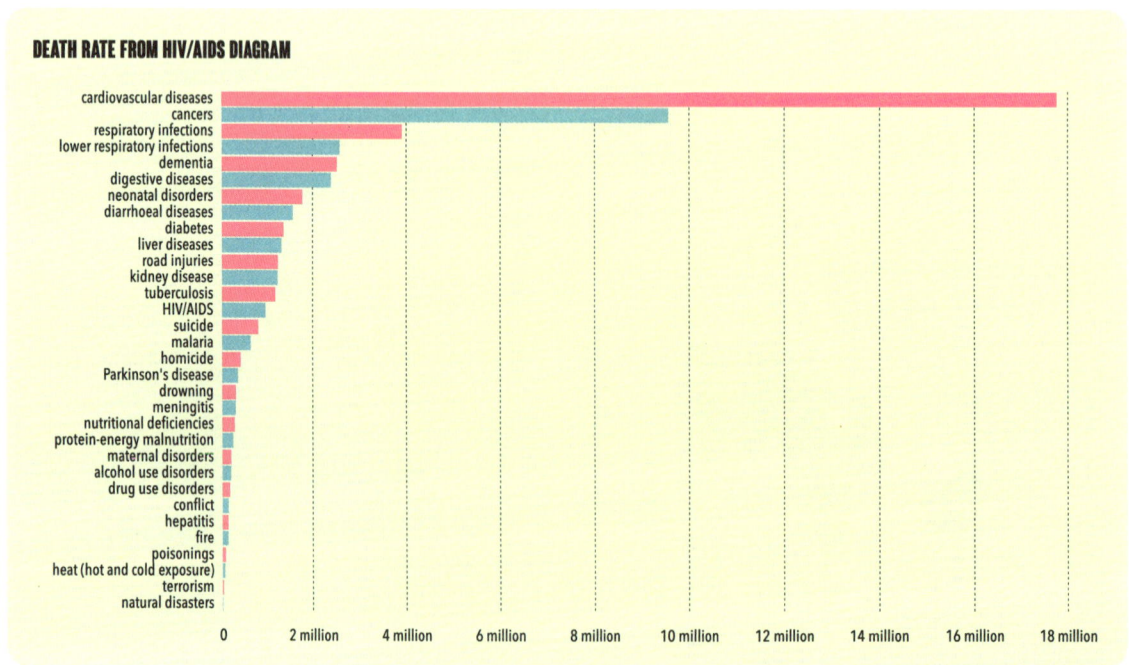

DEATH RATE FROM HIV/AIDS DIAGRAM

HUMAN POPULATION DENSITY

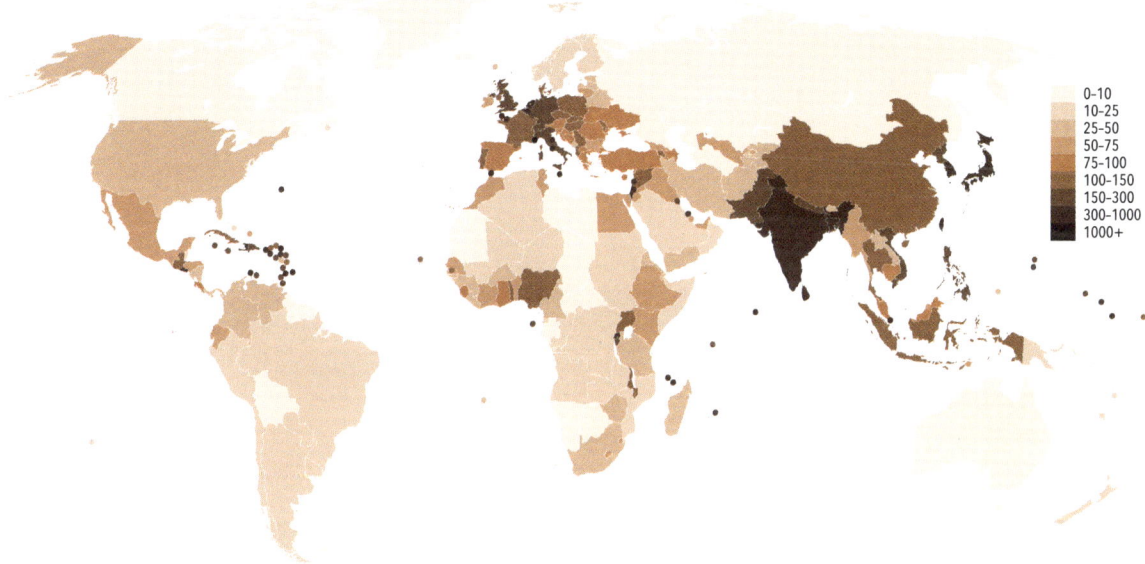

0-10
10-25
25-50
50-75
75-100
100-150
150-300
300-1000
1000+

NEW CHALLENGES

Recent decades have been marked by the appearance of new viral pathogens, such as the severe acute respiratory syndrome coronavirus (SARS-CoV), Middle East respiratory syndrome coronavirus (MERS-CoV), and severe acute respiratory syndrome coronavirus 2 (SARS-CoV-2), which was responsible for the worldwide pandemic of 2020. Many of these viruses bear strong similarity to viruses known to infect other species. For example, SARS-CoV-2 strongly resembles a bat-borne virus.

Such species-to-species transmission is probably traceable to a number of factors, including increasing encroachment of human communities on wildlife habitats, which forces animals to live in denser and denser populations; the movement of more and more human beings into densely packed cities, permitting higher rates of transmission of such diseases; and increasingly rapid travel patterns, which quickly disseminate diseases such as viral respiratory infections internationally.

As this brief account indicates, humanity has achieved great successes against some infectious diseases, particularly through prevention via improved diet, water supply, sanitation, and immunization. But the circumstances of human life have also changed in ways that promote the transmission of infectious diseases – most notably, increasing population density and range and speed of travel of infected individuals. The balance between pathogen and host keeps shifting.

Opposite: A micrograph of the HIV virus.
Above: Population density (people per km²) by country, 2006.
Below: An artist's impression of a coronavirus.
The novel coronavirus SARS-CoV-2 emerged in late 2019.

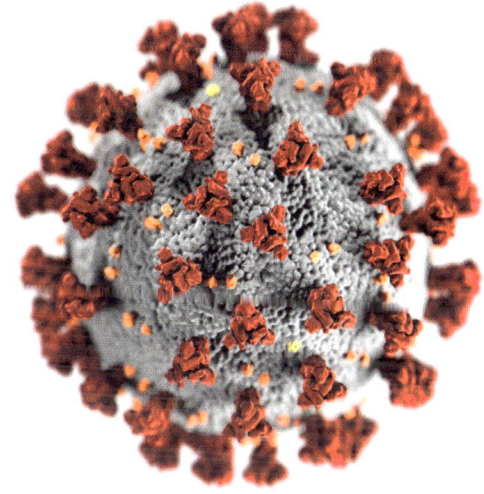

2

INFECTIOUS DISEASE

Human beings and the organisms that infect them have been locked in a dance of disease and death since before the dawn of recorded history. Prehistoric human remains show clear evidence of infectious disease, such as tuberculosis of the spine, and recorded history is studded with various plagues and the devastation they wreaked on individuals, cities, and whole civilizations. Even today, it is estimated that at least 10 million people die every year from infectious disease, most commonly pneumonia.

AGENTS OF INFECTION

Infectious agents can be helpfully divided into broad categories, each of which contains important subcategories: viruses, bacteria, fungi, parasites, and arthropods.

Viruses are one of the most remarkable infectious agents, in part because they do not live independently of the organisms they infect. Considered by itself, a virus is just genetic material, DNA or RNA, which codes for the proteins that make it up, as well as an outer shell consisting of proteins. Viruses are so small that they cannot be seen with a light microscope. Viral diseases include the common cold, influenza, and SARS (severe acute respiratory syndrome).

Bacteria are single-celled organisms, some of which cause disease but without which no animal could live. For example, some of the thousands of species of bacteria that inhabit the human gut produce vitamin B_{12}, and "good" bacteria in the intestines and on the skin help to keep disease-causing ones at bay. They are large enough to be seen with a microscope. Different bacteria cause diseases such as food poisoning, tuberculosis, and syphilis.

Bacteria are prokaryotes, meaning that they have no cell nucleus, but fungi are eukaryotes, indicating that their cells have nuclei and other membrane-bound organelles. They include organisms such as yeasts, moulds, and mushrooms. While some fungi cause human diseases including candidiasis, coccidioidomycosis, and histoplasmosis, they are also responsible for many antibiotics, the rising of bread, and the production of beer and wine.

Parasites come in many shapes and sizes. Some, such as the organism that causes malaria, are just single cells, while others, such as roundworms and tapeworms, are large enough to be seen with the naked eye. Parasites generally live within their host, cannot survive without it, and harm their host to some degree, in ways that range from stealing nutrients to causing death.

Arthropods are small organisms that have an external skeleton, a segmented body, and paired, jointed limbs. They include such familiar organisms as insects and spiders. When arthropods such as ticks and fleas transmit disease, their presence is typically referred to not as an infection but as an infestation. However, such infestations can transmit other diseases, such as bubonic plague.

Opposite, from top left: Congenital syphilis infecting a sweat duct; tapeworm parasites; *Candida albicans*, an opportunistic fungal pathogen that can infect those with HIV.

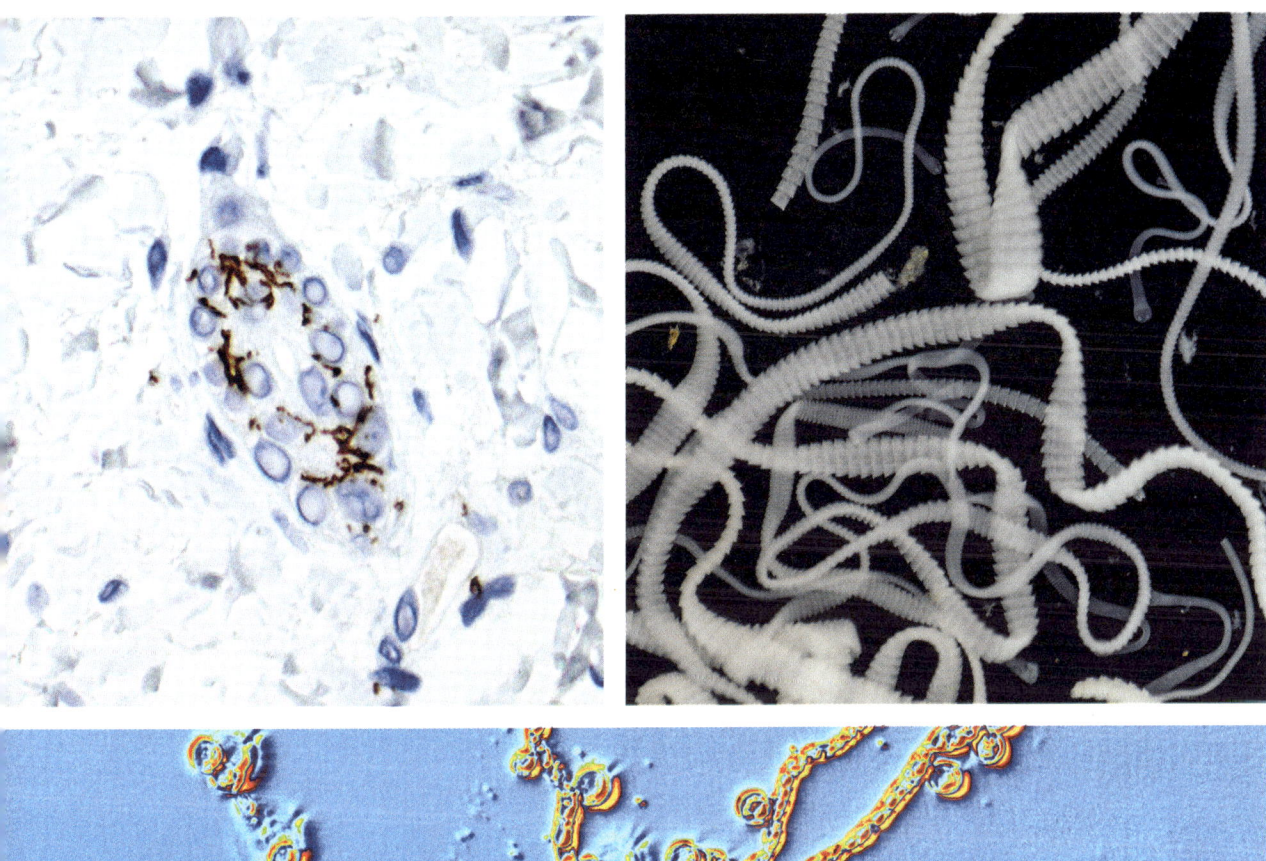

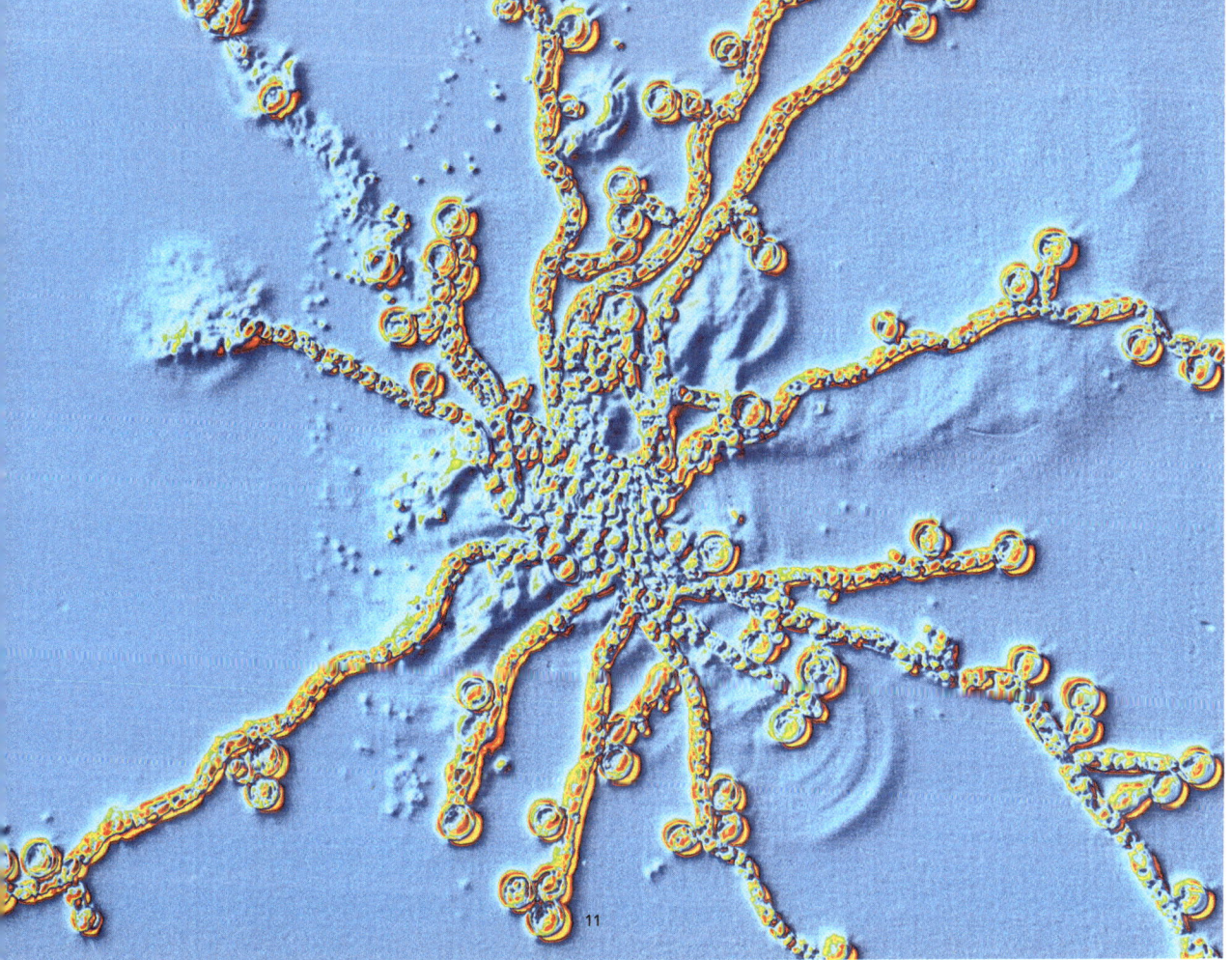

11

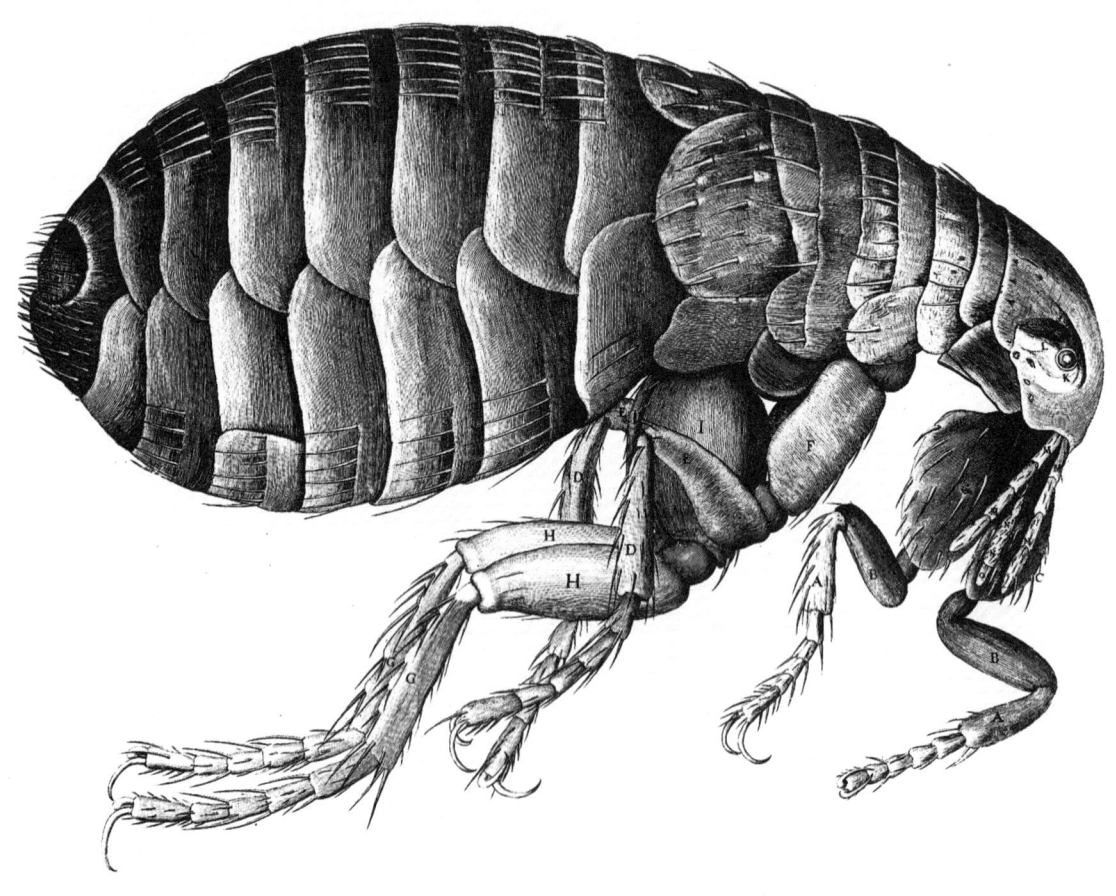

Above: Robert Hooke's famous drawing of a flea, from his *Micrographia*.

INFECTIOUS AGENTS ARE CALLED CONTAGIOUS WHEN THEY ARE EASILY TRANSMITTED FROM ONE PERSON TO ANOTHER.

DISEASE

Infectious agents are called contagious when they are easily transmitted from one person to another. Some are transmitted directly, through respiratory droplets or sexual contact. Others spread indirectly, through such routes as contaminated food and drinking water, non-human animals, and insect bites.

Physicians distinguish infections from many other forms of disease, including congenital disorders such as spina bifida, inflammatory disorders such as rheumatoid arthritis, traumatic injuries, metabolic diseases such as diabetes, benign and malignant tumours, and vascular disorders, such as heart disease and stroke. Yet it must be said that even some of these diseases can be traced to infection, such as the human papillomavirus's role in causing cervical cancer.

Infections can involve many different organ systems. The respiratory system represents the single most common portal of entry and site of the greatest number of cases of infectious diseases, such as the common cold and influenza. The digestive and urinary systems are other common sites of infection, including food poisoning and run-of-the-mill urinary tract infections. In fact, every organ in the human body, including the brain, heart, and bones, is subject to a variety of infectious diseases.

Mere exposure to an infectious agent does not necessarily produce disease. For example, some bacteria that normally live on the skin cause no harm unless they gain access to other body compartments, such as a joint. Some infectious agents are more virulent (more likely to cause disease) than others. Another key factor is host resistance to infection. Patients with impaired immune systems may develop disease from an exposure that would have no effect if the immune system were functioning normally.

DIAGNOSIS

Physicians diagnose infectious diseases by a variety of means. The most venerable are history and physical examination. Does the patient have typical symptoms of infection, such as fever, cough, vomiting or diarrhoea, or rash? Exposures are especially important. Has the patient been around some else with similar symptoms, or can a cluster of cases be traced to a particular source, such as contaminated food or drinking water?

A key technological development in the diagnosis of infectious disease was the introduction of the microscope. When tissues such as blood or bodily fluids such as sputum or urine are stained and examined with a light microscope, organisms such as bacteria and fungi are often visible. More advanced techniques, such as electron microscopes, are much more powerful and can visualize smaller infectious agents, such as viruses.

Another important diagnostic technique is cell culture. Bacteria and fungi can often be grown in a petri dish containing a growth medium, and their colonies often have a characteristic appearance. Infectious agents such as bacteria can then be subjected to different antibiotics, to determine which ones they are susceptible to. Some infectious agents, such as viruses, can only be cultured in other living organisms, such as the embryos in chicken eggs.

Other even more sophisticated tests look for specific molecules associated with specific infectious agents. When an infectious disease evokes the production of antibodies by the host organism, the presence of antibodies proves infection, as in a common test for strep throat. The presence of specific enzymes can be used to identify viruses. More recent tests based on the polymerase chain reaction (PCR) look for specific nucleic acids associated with an infectious agent.

THE LIFE OF AN INFECTIOUS MICROBE

Suppose you are an infectious microbe. You are very small. The largest human cell, an egg, released by the human ovary about once every 28 days in women of childbearing age, has a diameter of about 0.1 mm, meaning about 10 of you could line up end to end in the space of 1 mm. A typical bacterium, by contrast, is about 0.001 mm, only 1/100 as wide. You could line up about 1,000 bacteria in 1 mm. The diameter of a typical virus is about 0.0001 mm, meaning you could line up 10,000 viruses in 1 mm.

You are also incredibly numerous. It has been estimated that the number of bacteria on earth is about 5 million trillion trillion, which is the number 5 followed by 30 zeroes. This number exceeds the known number of stars in the universe. The number of viruses is estimated to be even more, about 10 to the 31st power. Compare these figures to the number of cells in an adult human body, which is estimated to be about 50 trillion, or the number 50 followed by 12 zeroes.

You are also incredibly pervasive. Bacteria are found throughout the earth, from the deepest depths of the ocean to as high as 40 miles up in the earth's atmosphere. Extremophiles, bacteria that thrive under extreme conditions, are found in such unlikely places as geothermal vents, where temperatures reach 121 °C, under Antarctic ice at temperatures of -25 °C, and at incredibly high levels of pressure and acidity. And viruses infect all forms of life, including plants and animals, as well as bacteria.

You are also, at least in many cases, incredibly intricate and beautiful. Many viruses, for example, exhibit an icosahedral shape with many component symmetries, including three-fold symmetries on their faces and five-fold rotational symmetries on their vertices. One attribute they lack, however, is colour – they are simply too small to be seen by visible light. Their beauty is even stranger because some viruses inflict terrible damage on the organisms they infect.

You are also incredibly busy. You share at least two functions with every other living organism – namely, survival and reproduction. To do so, you need to carry out a number of activities in sequence. You might think of these activities as a chain of six links, each link of which is necessary for the whole chain to remain intact. These are the very same six steps that physicians and scientists striving to prevent and cure infectious diseases seek to disrupt.

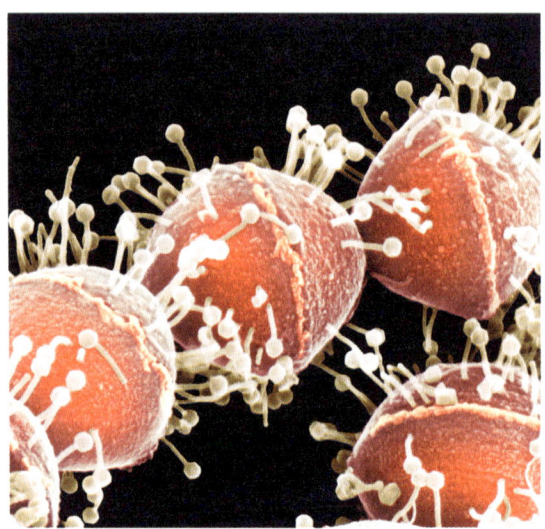

The Life of an Infectious Microbe

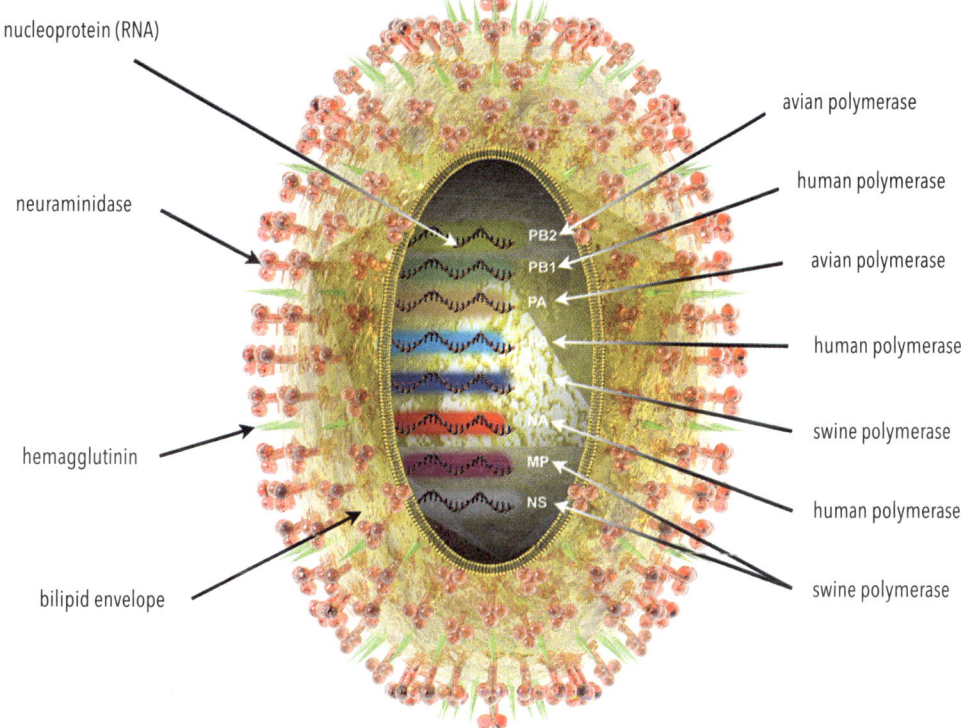

THE CHALLENGE

The first such task is to gain entry to a host. In the case of human diseases, the most common portal of entry for infectious organisms is the respiratory system, specifically the nose and throat, although other common sites include the gastrointestinal and urinary tracts. Many infectious microbes have adhesins on their surface that allow them to attach themselves to host cells. The cells to which they first attach are not always the ones in which they ultimately produce disease, however.

Your second mission is to establish yourself within a specific niche. Many pathogens contain or produce invasins, such as enzymes that enable them to gain access to host tissues and cells. Gaining entry is vital, because if the pathogen remains outside the cell, it generally lacks access to nutrients that the cell contains. For example, Salmonella and Shigella organisms, both of which cause diarrhoeal illnesses, enter the cells that line the intestines.

Your third mission is to avoid host defences. This is another reason that pathogens enter cells – once

Opposite: A virus (brown) attacking bacteria (red).
Above: The influenza virus H3N2, associated with seasonal flu.
Below: An illustration of a virus landing on a bacterium.

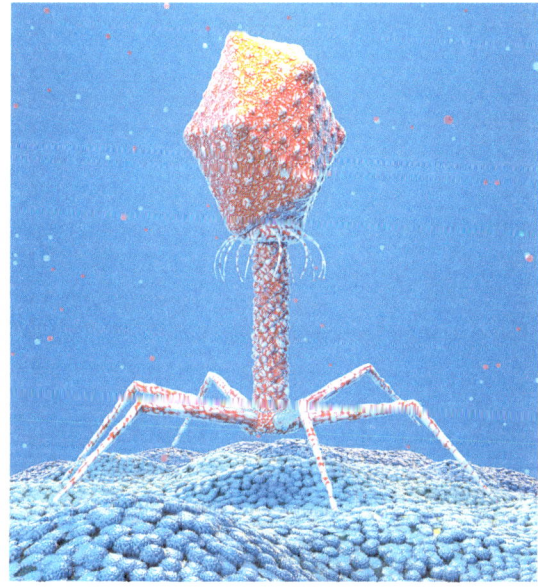

inside, they are often protected from host defences such as antibodies that circulate in the blood. But infectious microbes also avoid host defences while they are outside of cells. For example, some produce capsules that enable them to avoid phagocytosis, the process by which some white blood cells engulf and destroy bacteria.

Your fourth mission is to reproduce. One fascinating microbial reproductive strategy involves iron, which plays an important role in energy metabolism for both pathogens and host cells. Some bacteria and fungi produce molecules called siderophores (from "iron bearers"), which bind iron more avidly than host cells. In effect, the pathogen has adapted a means of "stealing" iron from its host, thereby providing it with a key nutrient necessary for survival and reproduction.

Your fifth mission is to produce toxins. After all, if you weren't somehow toxic or harmful to the host, you would not be a pathogen. Some pathogens produce exotoxins, molecules that damage host cells apart from the pathogen. One example would be the cholera bacillus, whose exotoxin causes severe diarrhoea. Endotoxins, by contrast, are on the surface of infectious microbes. One way endotoxins cause disease is by inciting a severe immune response, which can damage host cells.

Of course, as a pathogen, you want to be circumspect about your toxicity. If you cause too much damage to your host, you destroy your own home, perhaps before you have had time to reproduce sufficiently, or equally problematic, if your host hasn't been able to reproduce, in which case your progeny have no home. For this reason, many pathogens have developed mechanisms to modulate toxin production. Hosts get sick, but not too sick.

Your final mission is to survive long-term. This may mean taking up permanent residence in the host or transmitting your progeny to other hosts. An example of a pathogen that takes up permanent residence is the virus that causes chickenpox, typically a relatively harmless infection of childhood. After the acute infection passes, the virus resides in nerve cells near the spinal cord. Decades later, it can reactivate, causing the painful rash known as shingles.

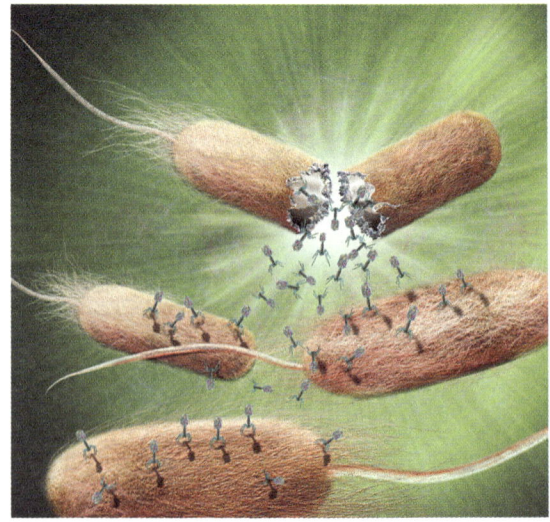

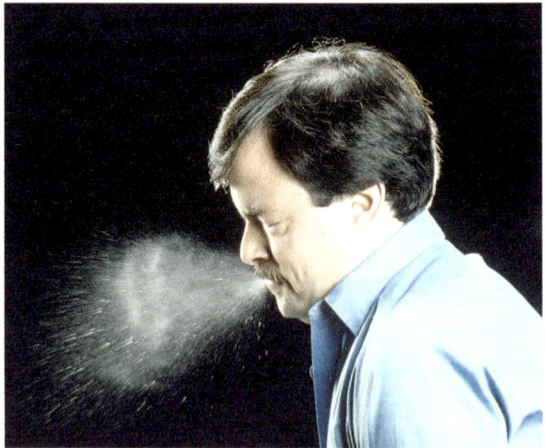

More familiar is the strategy of transmission to new hosts. Many respiratory viruses, such as the common cold, influenza, and coronaviruses, are transmitted through respiratory droplets, for example when a patient sneezes or coughs. Others, such as cholera, are transmitted by a faecal–oral route, which is amplified by severe diarrhoea. Still others, such as the retrovirus that causes HIV/AIDS, can be transmitted through sexual contact or by sharing needles.

Above: A virus uses the host's genetic material to replicate itself. Viruses are frequently spread through aerosolized contagion.
Opposite: Detailed micrograph of virus particles invading and destroying a bacterium.

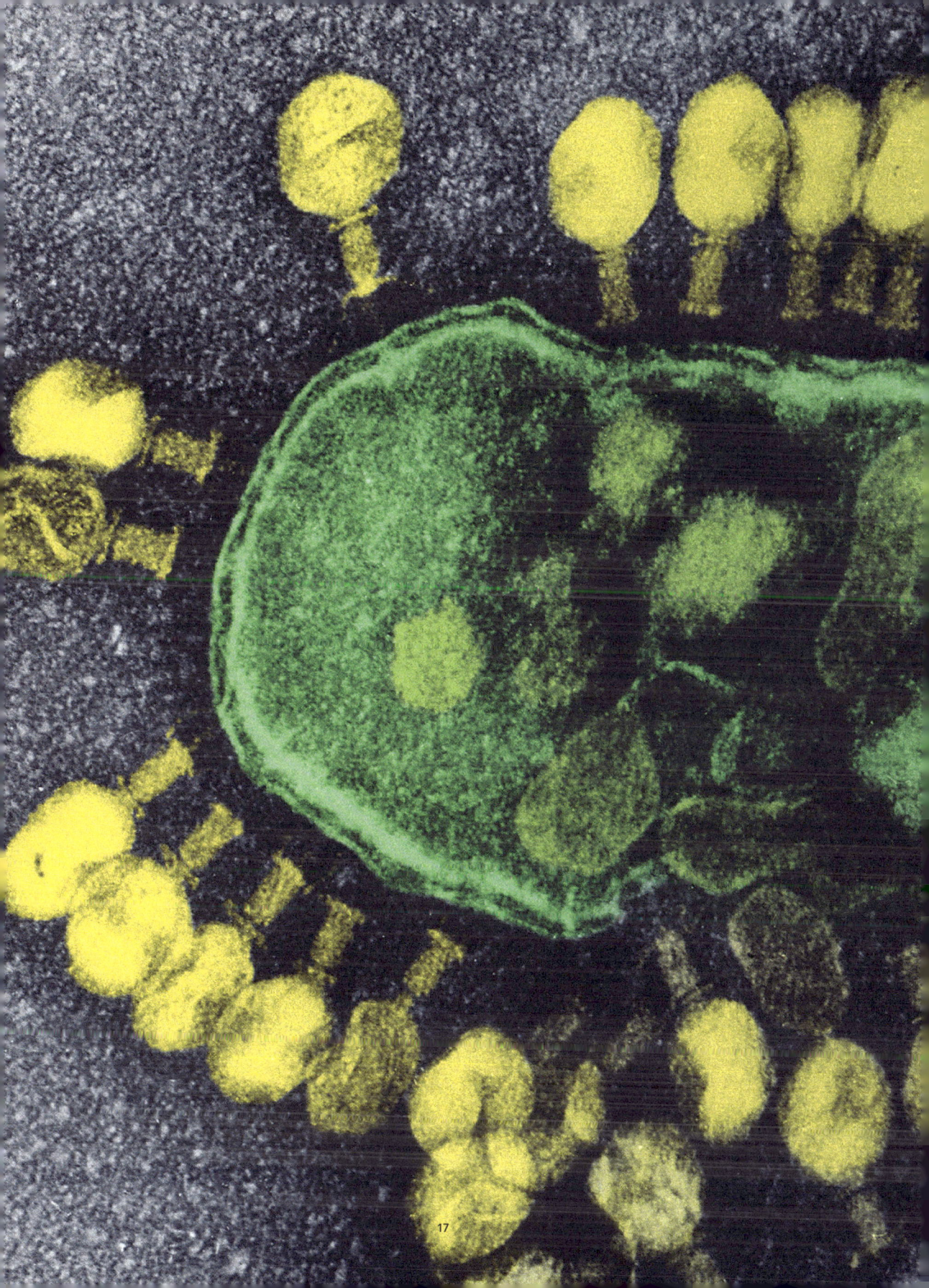

NATURAL SELECTION AND INFECTIOUS DISEASE

The biosphere is not an entirely hospitable environment. That this would be the case is predicted by the theory of natural selection, as outlined by Charles Darwin (1809–82) and others, and subsequently refined by geneticists. Darwin's theory combined several observations:

1. Individual members of a species vary from one another in many ways. In the human case, for example, some people are taller than others.
2. The characteristics of some individuals give them an advantage in survival and reproduction over other individuals. If being able to pick fruit from trees were advantageous, tall people might tend to survive and reproduce more than short people.
3. Conversely, those who are less well adapted to their environment would be at a disadvantage. Short people, for example, might not be able to pick as much fruit.
4. Over time, the frequency of a particular characteristic will tend to change. For example, assuming height can be passed from generation to generation, average height will tend to increase.

SPREAD OF HOMO SAPIENS OVER FACE OF EARTH OVER TIME (YEARS)

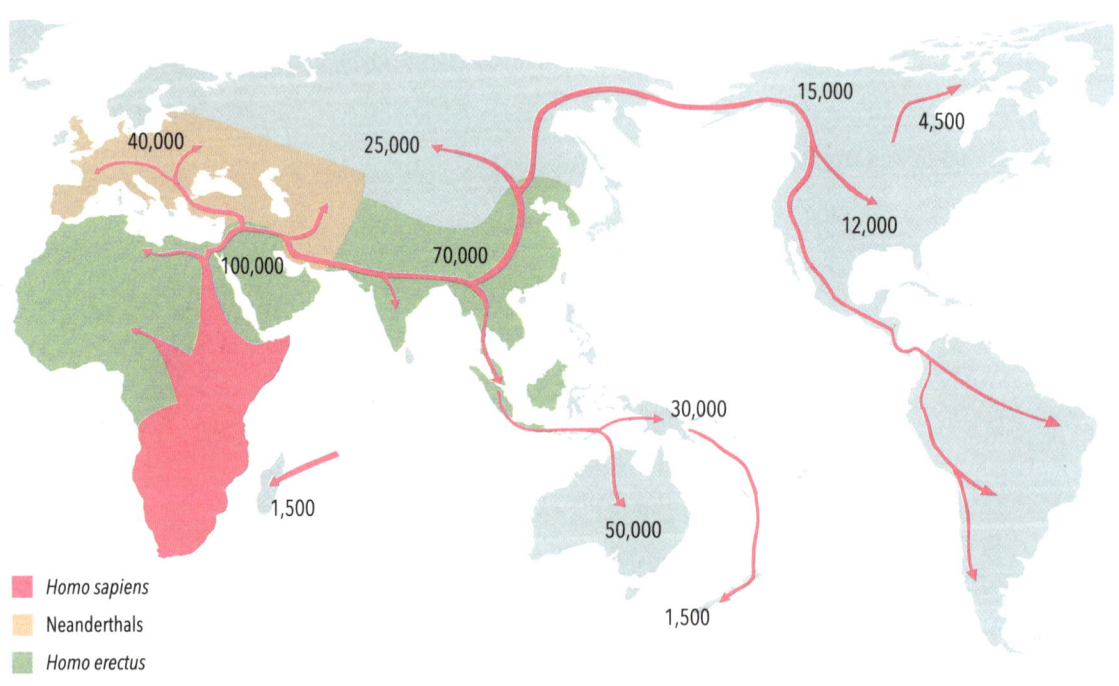

Just as height might be selected for in a human population, various characteristics might also undergo natural selection in microbial populations, which may in turn influence the characteristics of human populations with which they coexist. And the same applies in reverse – many changes in the patterns of human life have taken place over tens of thousands of years, which have created new challenges and opportunities for microbes.

Human beings have spread out over much of the surface of the earth, encountering new microbes and animal species that harbour them, such as birds and bats. Humans have also domesticated many animal species, such as dogs and cattle, living in closer contact with them. Finally, the density of human populations has increased with transitions in societies from hunter-gatherer to agricultural to industrialized and urban models, creating new opportunities for microbial transmission.

Above: Charles Darwin (1809–82).
Right: Alexander Fleming (1881–1955), who discovered penicillin.
Below: Fleming's accidental discovery was the world's first antibiotic.

ANTIBIOTIC RESISTANCE

Consider, for example, the phenomenon of antibiotic resistance. With the discovery of the antibiotic penicillin by Alexander Fleming (1881–1955) in 1928, humans attained a powerful new weapon against pathogenic bacteria, and in the ensuing decades many additional antibiotics were added to medicine's antibacterial armamentarium. Some speculated that these new medications might spell the end of certain types of infectious disease.

But as humans were introducing new types of antibiotics, microbes were not sitting still. Because bacteria are so numerous and undergo genetic mutations at such a relatively high rate, occasional bacteria express resistance to an antibiotic. Bacteria may have been resistant in the first place – for example, many gram negative (unaltered by a particular stain) bacteria are naturally resistant to penicillin. In other cases, mutation leads to resistance. And in still other cases, one species may acquire resistance from another.

Bacterial resistance can take many forms. For example, the bacterium can produce a molecule that inactivates the antibiotic, such as an enzyme that breaks it up. Second, it can alter a metabolic pathway

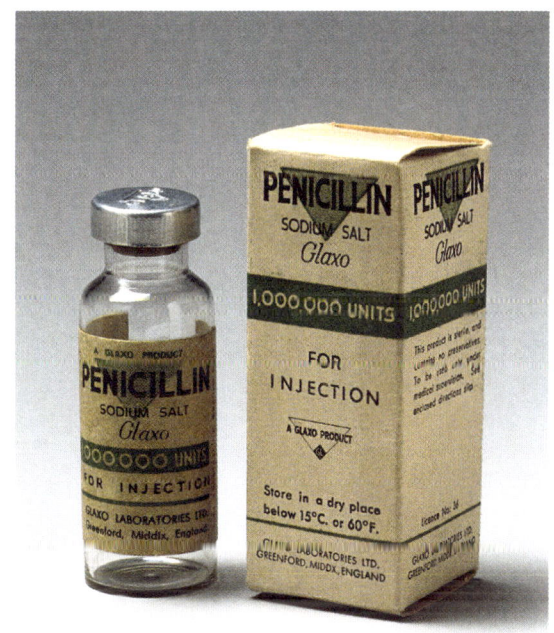

so that it no longer depends on a substance that the antibiotic blocked. Third, it can alter a binding site so that the drug no longer attaches to it. And fourth, a bacterial cell can actively pump the drug out of itself, preventing it from reaching sufficiently high concentration to work.

Natural selection explains how this happens. Suppose a new type of antibiotic is introduced that kills a certain type of pathogenic bacterium. As the drug is administered to many patients, almost all the bacteria die. But suppose a few bacteria bear a mutation that makes them resistant. As all the susceptible bacteria are wiped out, reducing competition for nutrients and other resources, these few resistant bacteria begin to multiply and are soon causing disease all over again.

And this occurs not only among bacteria, but in other pathogenic species, as well. For example, fungi can develop resistance to antifungal drugs, protozoa can develop antiprotozoal resistance, and viruses can develop antiviral resistance. The longer the period that infectious microbes are exposed to a medication,

Below: Sickle cells blocking a blood vessel in a sufferer from sickle cell anaemia.
Opposite: A page from Alexander Fleming's notebook, showing the petri dish where the penicillin was grown.

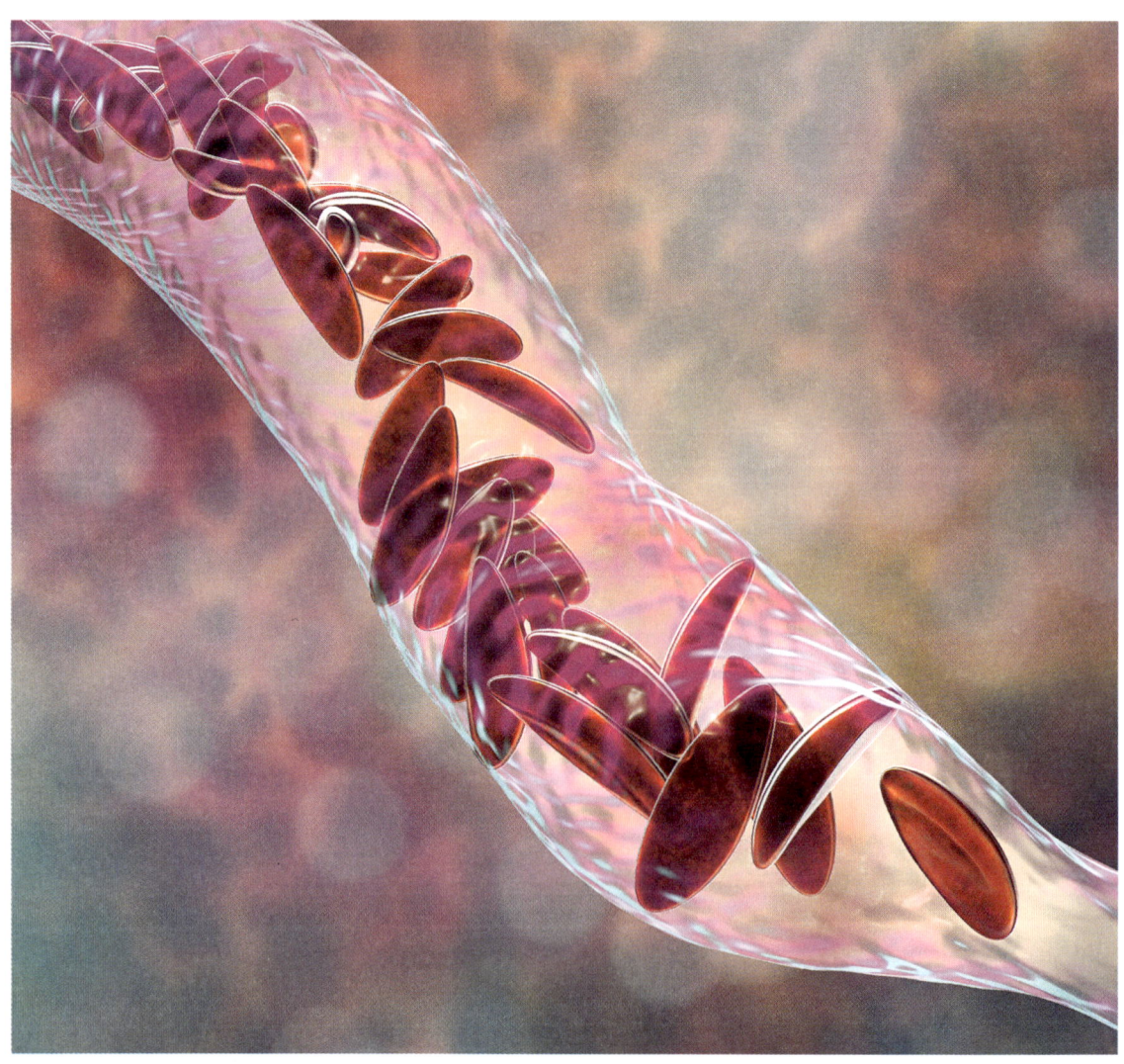

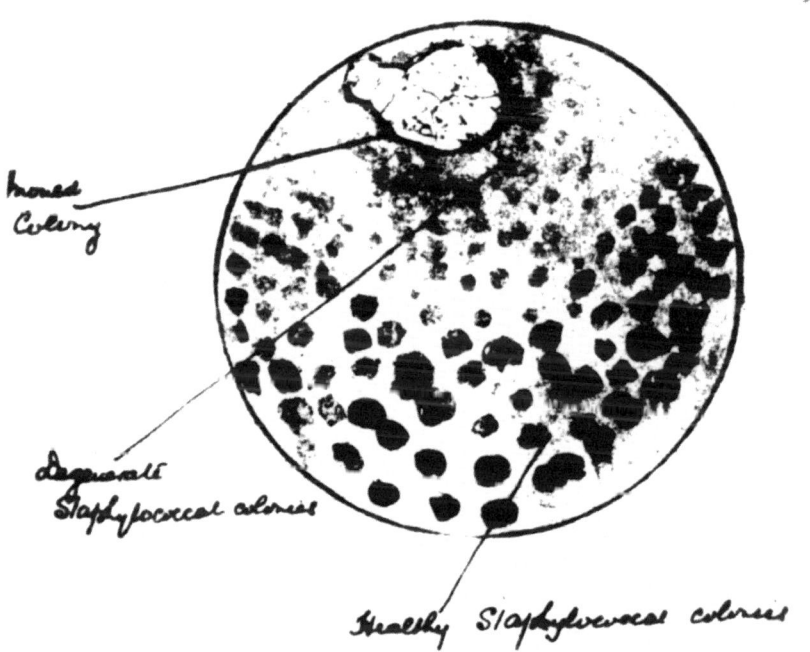

Anti-bacterial action of a mould (Penicillium notatum)

Labels: Mould colony; Degenerate Staphylococcal colonies; Healthy Staphylococcal colonies

On a plate planted with Staphylococci a colony of a mould appeared. After about two weeks it was seen that the colonies of Staphylococci near the mould colony were degenerate.

SELECTION FOR ANTIBIOTIC-RESISTANCE AMONG BACTERIA

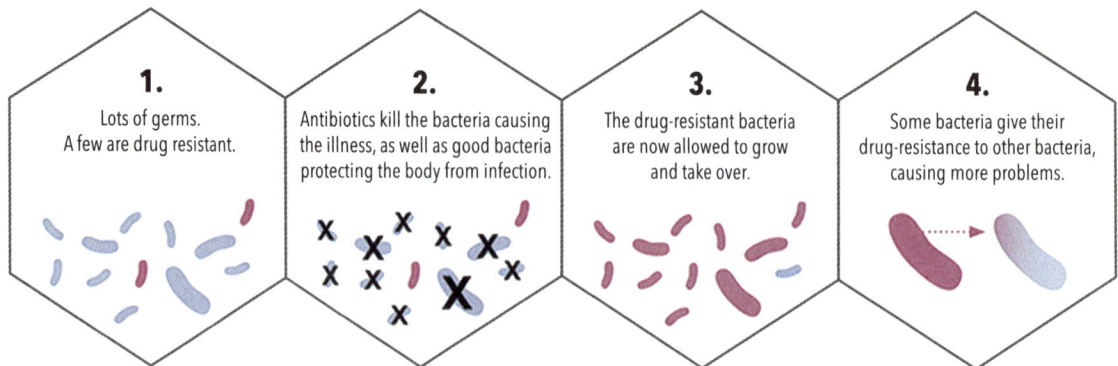

the greater the selective pressure for the growth of resistant organisms. The microbes are not "trying" to beat the drugs, but the drugs are selecting for resistant microbes.

COUNTERING ANTIBIOTIC RESISTANCE

There are a number of ways to counteract antibiotic resistance. One is never to take antibiotics unnecessarily, which selects for resistance while providing no benefit. It is estimated that half or more of antibiotic prescriptions are inappropriate, and the problem is compounded by the fact that some patients self-medicate without understanding this risk. Another major problem is the widespread use of antibiotics in livestock to promote growth, which also selects for resistant organisms.

HOST RESISTANCE TO PATHOGENS

Of course, natural selection isn't working only on microorganisms. It is also at work on their human hosts. One example of host resistance to a pathogen are mutations in the haemoglobin molecule, which enables red blood cells to transport oxygen, in areas where malaria is endemic. Sickle cell disease results when a patient inherits abnormal haemoglobin genes from both parents, which causes red blood cells to assume an abnormal shape, shortening both the cell's and the patient's lifespan.

However, when individuals have only one copy of the abnormal gene (sickle cell trait), they enjoy increased resistance to malaria, the mosquito-borne disease associated with recurrent fevers that can prove fatal. Not surprisingly, the sickle cell gene is found most commonly among people whose ancestors lived in areas where malaria is endemic, especially sub-Saharan Africa and around portions of the Mediterranean. Other genetic traits, such as thalassemia, are also protective.

Human beings and the microbes that infect them are engaged in a kind of arms race, selecting for certain genetic variations in both populations. When a pathogenic microbe expresses a new characteristic that poses a greater threat to its host, there is greater selective pressure on the host population for a resistance factor. Likewise, when humans introduce a new preventive strategy or therapy for an infectious disease, there is increased pressure on the microbe for a variant that can circumvent it.

Yet it would be a mistake to assume that either party, host or pathogen, is simply trying to eradicate the other. As we have seen, a super-bug that rapidly killed every human being it infected would quickly eat itself out of house and home, thus undermining its own prospects for survival. Likewise, whenever humans attack one microorganism, they shift the balance between many microorganisms, which can have unintended and even unforeseen adverse consequences.

DISTRIBUTION OF MALARIA AND DISTRIBUTION OF SICKLE CELL GENE

a

Sickle haemoglobin (HbS) data points
- Presence
- Absence

b

HbS allele frequency (%)
- 0–0.51
- 0.52–2.02
- 2.03–4.04
- 4.05–6.06
- 6.07–8.08
- 8.09–9.60
- 9.61–11.11
- 11.12–12.63
- 12.64–14.65
- 14.66–18.18

c

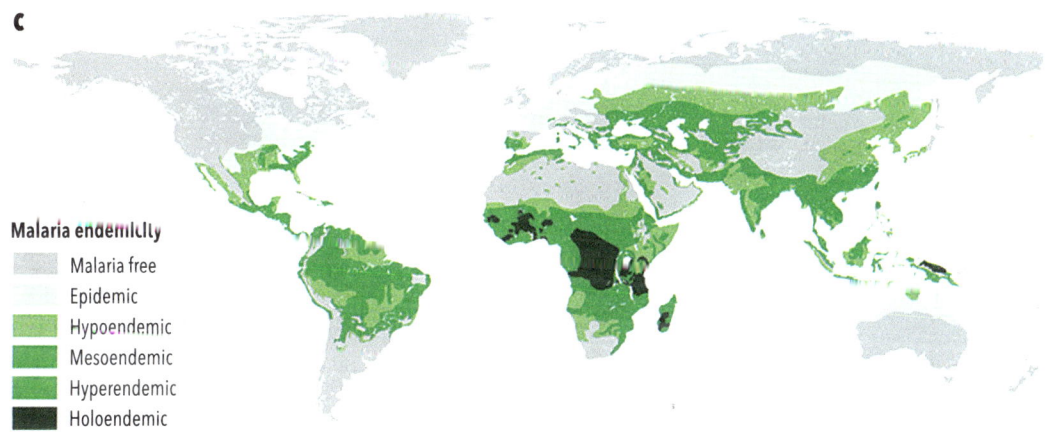

Malaria endemicity
- Malaria free
- Epidemic
- Hypoendemic
- Mesoendemic
- Hyperendemic
- Holoendemic

5 ANCIENT VIEWS OF HEALTH AND DISEASE: THE HIPPOCRATICS

To know where we are and where we are headed, it is helpful to know where we have come from. Some ideas about health and disease that today we would regard as preposterous were once dogma, and some of our contemporary ideas would have been dismissed as lunacy in past times. Likewise, some of the notions that we treat as axiomatic today may someday be regarded as preposterous, and those who follow us may in the future look back at our medicine and shake their heads in astonishment, wondering how we could have been so blind.

There are problems with approaching the history of medicine as a largely biographical affair, such as the tendency to attribute too much influence to single individuals and to overlook the contributions of teams, communities, and culture that develop over multiple generations. Nevertheless, there have been individuals in medicine's history who, though perhaps not solely responsible for the ideas and practices we attribute to them, nevertheless embody a view of health and disease that deserves serious consideration. One such individual is the ancient Greek physician Hippocrates (460–c. 370 BC).

A HOLISTIC APPROACH

One of the most remarkable features of Hippocratic medicine, at least compared to medical practice today, is its holism. To understand a person's state of health, Hippocrates believed, it is necessary to look at the whole person, including not only the whole body but the person's way of life and how that mode of living fits into the larger context of a person's environment. The Hippocratic Oath, though perhaps not penned by Hippocrates, even makes explicit the suggestion that physicians should see patients at home, rather than in the artificial environment of an office or hospital.

HIPPOCRATES

The details of the life of Hippocrates are at best sketchy. He seems to have been born on the Greek island of Kos, the son and grandson of physicians, and whose own sons would follow family tradition and practise medicine themselves. Plato mentions Hippocrates in his dialogues, asserting that a physician must understand the body in order to practise medicine well. Aristotle also refers to him, calling him "the Great Hippocrates". So, while many biographical details are missing, we can be confident that Hippocrates was widely regarded in his day as a paragon of medical excellence.

Ancient Views of Health and Disease: the Hippocratics

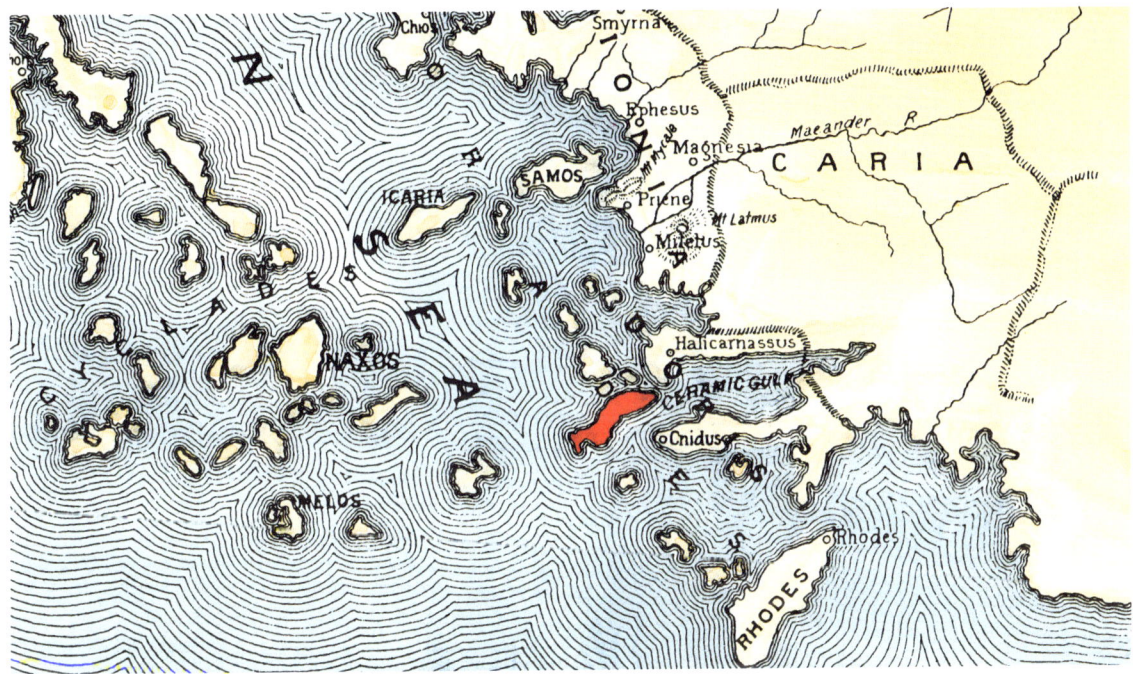

To see patients outside the context in which they live would necessarily distort the physician's understanding of their state of health. Today we might say that people exist in habitats or ecosystems, and that their relationship with their environment shapes whether they will be healthy or ill. Does the patient live alone or in a large family? Is the patient rich or poor? Is the patient's home life harmonious or full of strife? What are the patient's customary diet, exercise practices, and sleep patterns? The answers to such questions determine the prevention of disease and promotion of health.

THE MODERN-DAY APPROACH

Today, by contrast, physicians often proceed by analysis, which literally means a cutting up. For example, one of the first rites of medical school is to begin dissecting a cadaver. To understand the human body, we believe, we need to divide it up into its component parts. The Greeks, by contrast, eschewed dissection. When a patient presents with symptoms, we attempt to localize disease. Supposing that the patient has pain in the right upper part of the abdomen, we ask ourselves, could it be the liver, the gallbladder, the duodenum, right kidney, and so on?

THE HUMORAL MODEL

The Hippocratics operated with a primarily humoral model of health and disease. The question was not, where is the seat of the disease in the body – which organ, tissue, or cells? Instead, the key step was to determine which forces in the patient's life had got out of balance. The humours include yellow bile, associated with fire, black bile, associated with earth; phlegm, associated with water; and blood, associated with air. When a patient developed a fever, for example, it suggested an excess of heat, which was associated

Above: Hippocrates was born on the island of Kos (in red), in the southeast of the Aegean Sea.

> **ONE OF THE MOST REMARKABLE FEATURES OF HIPPOCRATIC MEDICINE, AT LEAST COMPARED TO MEDICAL PRACTICE TODAY, IS ITS HOLISM.**

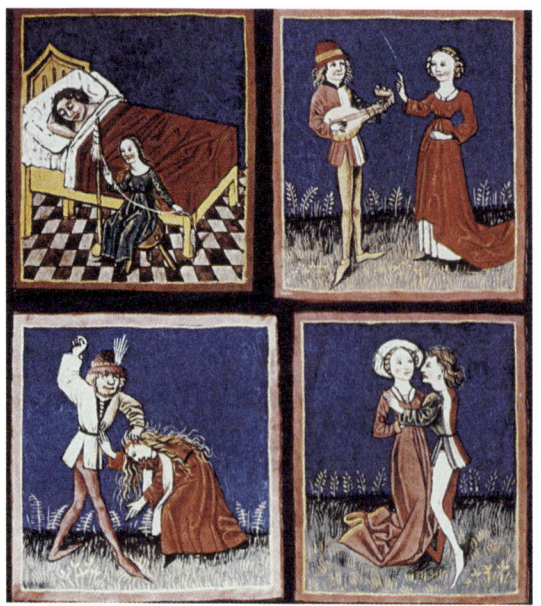

with air and fire, and could therefore be treated by bleeding to release excess heat.

The humoral theory placed a premium on the maintenance of balance. The force of each of the humours could be a good thing, if each were kept in balance, but if one humour became excessive or deficient, imbalance and illness would result. Inputs such as food and drink needed to be balanced against outputs such as urine and sweat. If inputs were excessive, balance could be restored by ordering the patient to fast or administering purgatives. This Hippocratic way of thinking persists in the homespun advice to starve a fever and feed a cold.

THE HIPPOCRATIC LEGACY

Lest we suppose the humoral model a mere anachronism, we should notice that even today, in an era of highly analytical medicine, we sometimes still think in recognizably Hippocratic terms. For example, we recognize that many humoral substances, such as electrolytes including potassium and sodium, gases such as oxygen and carbon dioxide, and hormones such as thyroid hormone, need to be kept within a fairly tight range for patients to remain healthy. Both excess and deficiency can each cause and serve as an indicator of illness.

SELF-HEALING VS. GERM THEORY

The Hippocratics operated with great faith in the human organism's ability to heal itself. The physician's task, once a diagnosis was reached, was to remove impediments to this natural tendency toward health. For example, if the patient's illness could be traced to an excess of cold and damp, the patient might be counselled to move to a warmer, drier climate. Nor was the Hippocratic physician ever treating the disease itself – diagnosis and therapy were always directed at the patient. The aim was not to eradicate disease but to help the patient return to a natural state of health.

These ideas held sway for over two millennia, persisting until the nineteenth century. Those promoting the germ theory as an approach to what we now know as infectious disease were fighting an uphill battle. For example, the notion that a single type of organism – and one too small to be visible, at that! – could cause serious disease might have struck the Hippocratics as laughable. How could a single factor such as a type of bacterium disrupt health, and how could a single medicine targeting that factor possibly restore it? The Hippocratics took for granted that the whole patient comes first.

And in important respects, the Hippocratics were right. For example, we now know that one of the most effective means of preventing infectious disease is providing clean drinking water and disposing properly of sewage – a clearly environmental approach. Moreover, the zest for killing pathogens has sometimes produced diseases of its own. For example, when bacterial infections are treated with antibiotics, collateral damage to "good" intestinal bacteria can clear the way for "bad" bacteria, such as *Clostridium difficile*, to take up residence, causing a more serious and life-threatening infection.

Today notably Hippocratic perspectives seem to underlie high levels of interest in the gut microbiome. When *C. difficile* infections are treated with antibiotics, recent success rates have averaged only about 30 per cent. But when researchers attempted faecal transplants, essentially transplanting "good" bacteria from the intestines of healthy patients into the intestines of patients with *C. difficile* infections, success rates in some studies climbed to over 90 per cent. The key to fixing this bacterial problem, it appears, is not eradicating bacteria, but restoring a more normal balance of bacterial species in the gut.

So the Hippocratic spirit lives on. To be sure, it is now accompanied by other theories unknown to the Hippocratics, such as the anatomic localization of disease and the germ theory. In many ways, the Hippocratics are still in the ascendancy. While my CT scanner, prescription pad, and scalpel might indeed save your life, or at least restore you to health, it is also important to recognize that prevention of diseases such as heart disease, cancer, and stroke remain by far the best solution medicine has to offer. Many physicians I know don't go to the doctor very often, but they take promoting their health very seriously.

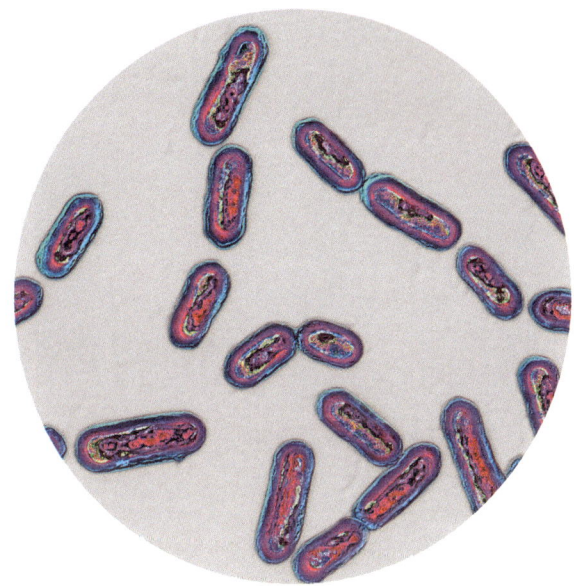

ANATOMIC LOCALIZATION OF DISEASE

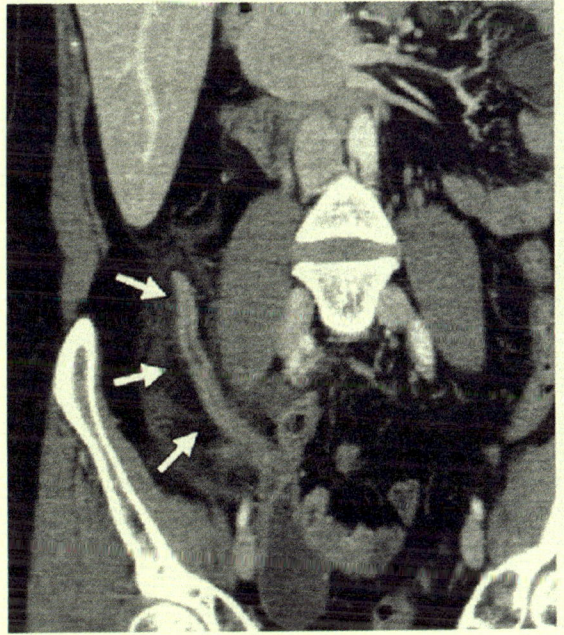

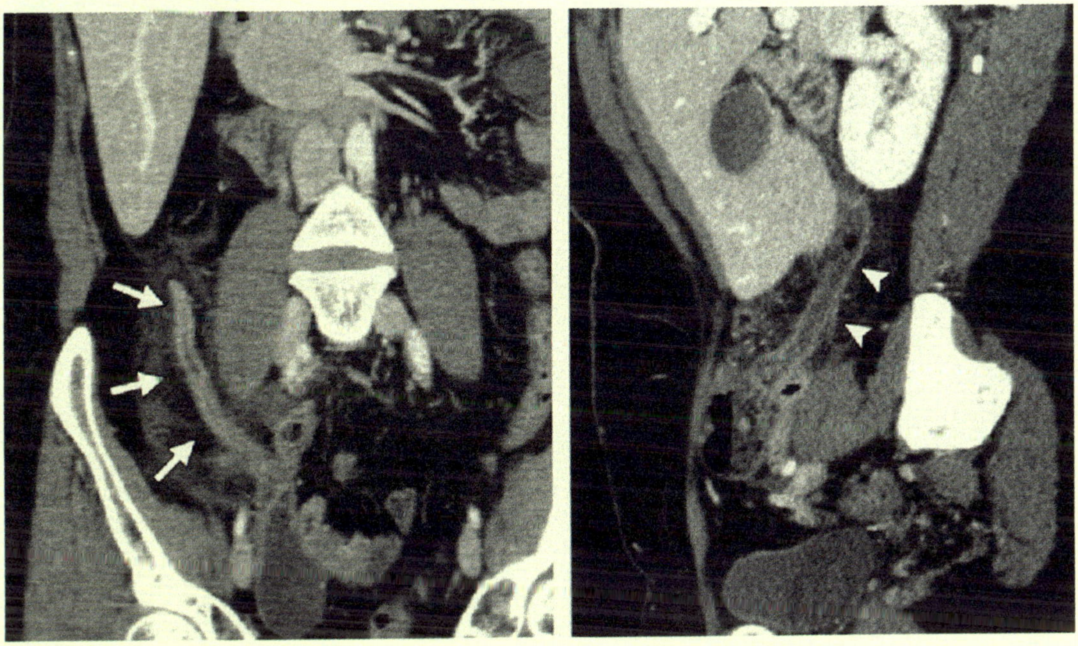

CT scan images showing an inflamed appendix (arrows) in a patient with appendicitis

Opposite: A medieval representation of the four humours. Clockwise from top: black bile (melancholy); blood (sanguine); phlegm (phlegmatic); yellow bile (choleric).

Above, top: *Clostridium difficile*, a bacterium associated with diarrhoea.

THE PLAGUE OF ATHENS

To understand infectious disease's power to alter the course of history, consider the plague in Athens.

PLAGUE AND THE PELOPONNESIAN WAR

The seeds of the Plague of Athens were sown by the Peloponnesian War (431–404 BC), which was described in detail by Thucydides. The Greeks lived in city-states of tens or hundreds of thousands of people, which entered into alliances with one another. As the power of seafaring Athens and its allies grew into an empire, Sparta and its allies felt increasingly threatened.

Sparta convened a conference of its allies in 432 BC, determining that if it continued to sit on the sidelines while Athens consolidated its power and influence throughout the region, it would end up isolated and weak. So the Spartans declared war. This spawned a classic contrast in approaches to warfare: the Spartans were the masters of land-based warfare, while the Athenians wielded greater naval power.

The Athenian strategy, as shaped by Pericles, was to avoid engaging Spartan troops in the field and instead rely on the unmatched Athenian fleet. They would lose the produce of the land around the city, but they could continue to rely on trade to feed their populace and maintain their economic might. To prevail, the Athenians just needed to remain patient and adhere to their defensive strategy.

Ironically, the Athenian strategy also sowed the seeds of plague, which first appeared in 430 BC. As people from the surrounding countryside moved within the city's walls, what had already been

a large city of perhaps one hundred thousand people quickly multiplied in size, becoming a terribly overcrowded one. Shortages of resources developed, inhabitants lived in close quarters, and sanitation and hygiene suffered.

The plague may have entered the city through its port. As described by Thucydides, features and signs of the disease included headaches, fever, sore throat, cough, vomiting, diarrhoea, insomnia, and death. Modern scholars debate the identity of the pathogen, but leading hypotheses include typhus, typhoid, and a haemorrhagic fever such as Ebola.

The Plague of Athens

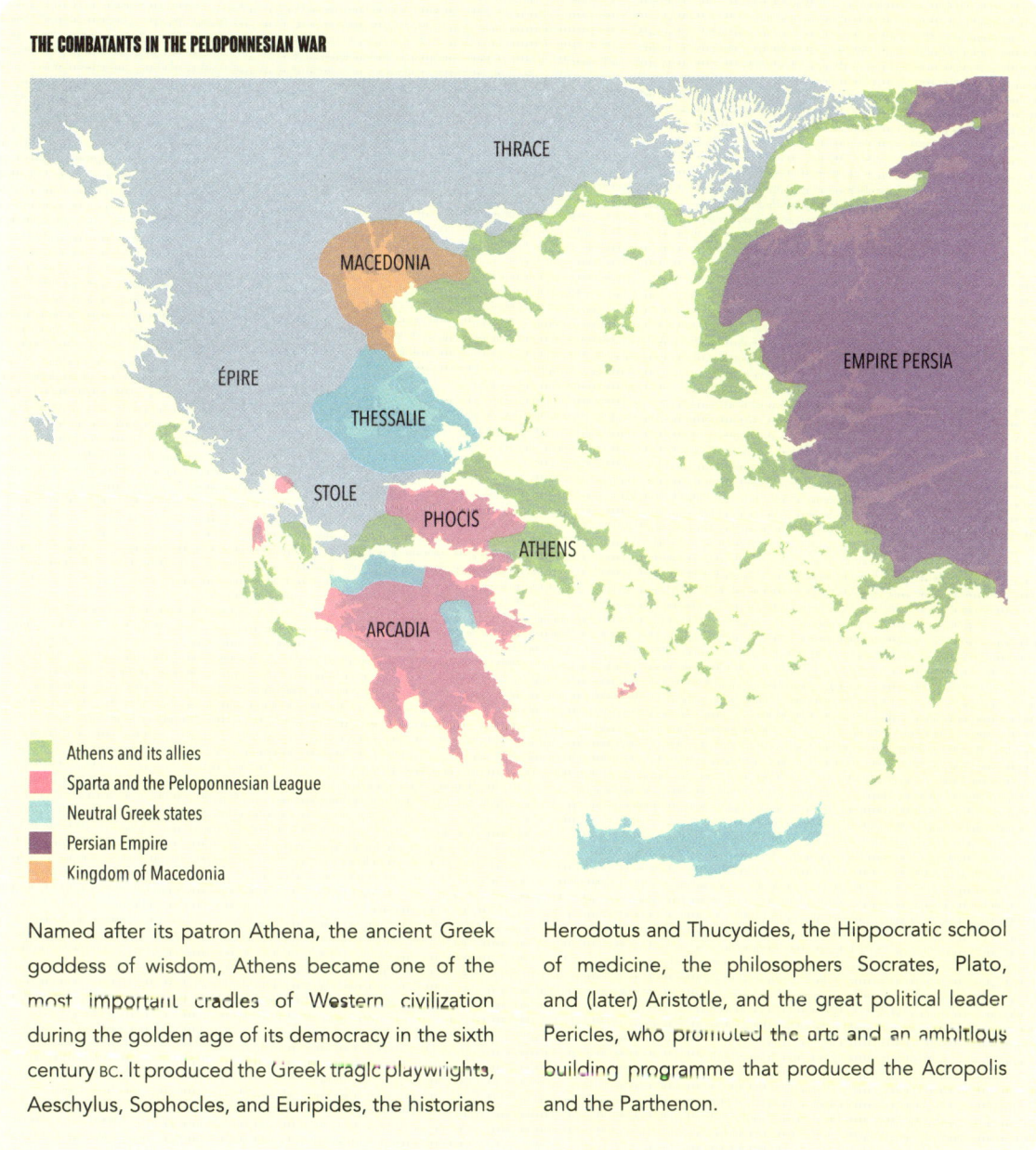

THE COMBATANTS IN THE PELOPONNESIAN WAR

- Athens and its allies
- Sparta and the Peloponnesian League
- Neutral Greek states
- Persian Empire
- Kingdom of Macedonia

Named after its patron Athena, the ancient Greek goddess of wisdom, Athens became one of the most important cradles of Western civilization during the golden age of its democracy in the sixth century BC. It produced the Greek tragic playwrights, Aeschylus, Sophocles, and Euripides, the historians Herodotus and Thucydides, the Hippocratic school of medicine, the philosophers Socrates, Plato, and (later) Aristotle, and the great political leader Pericles, who promoted the arts and an ambitious building programme that produced the Acropolis and the Parthenon.

It is estimated that the plague killed approximately 25 per cent of the population, perhaps as many as 25,000. Among those who died were Pericles, his wife, and their two sons. The smoke from the funeral pyres of the dead was so imposing to the Spartans that they withdrew their troops, for fear of contracting the disease. Thucydides, too, fell ill, but recovered and wrote his history.

Thucydides described the plague's devastation as "a kind of sickness which far surmounted all expression of words and exceeded human nature in its cruelty". It was

Opposite, from left: Thucydides told the story of the Peloponnesian War, along with a description of the plague, which used evidence-based Hippocratic theory. This modern reimagining shows Hippocrates directing the response to the plague.

impossible to predict who it would carry away, as "no difference of body, for strength or weakness, was able to resist it". And medicine, he wrote, was worse than useless, doing "good to no one but harm to others".

Worst of all, Thucydides wrote:

[w]as the dejection of mind in such as found themselves beginning to be sick (for they grew presently desperate and gave themselves over without making any resistance), as also their dying thus like sheep, infected by mutual visitation, for the greatest mortality proceeded that way. For if men forbore to visit them for fear, then they died forlorn; whereby many families became empty for want of such as should take care of them. If they forbore not, then they died themselves, and principally the most honest men.

SOCIAL IMPACT

The effects of the plague extended far beyond the biological. Thucydides describes "dying men tumbling upon one another in the streets, and men half-dead about every conduit through desire of water". This and the fact that the temples were full of dying and dead led people to "grow careless of both holy and profane things alike", producing a great licentiousness:

Neither the fear of the gods nor laws of men awed any man, the former because they concluded it was alike to worship or not worship from seeing that alike they all perished, nor the latter because no man expected that lives would last till he received punishment of his crimes by judgment. But they thought there was now over their heads some far greater judgment decreed against them before which fell, they thought to enjoy some little part of their lives.

In other words, no threat of punishment could deter those who already felt they were living under a death sentence. And why invest for the future, either property or honour, when no one expected to live long enough to enjoy their fruits? Those who conducted themselves most honourably by caring for the sick seemed to be most likely to contract the disease.

Thus the plague not only took the lives of a vast number of people, but it also precipitated a collapse of

Athenian society. Athens did not immediately succumb to the Spartans, but it emerged from the plague seriously weakened. Its material and human might had been reduced, and a general weakening of morale had also taken place, from which it would never regain its status as a major power.

The war carried on for many more years, and the Athenians might yet have emerged victorious. But they ignored Pericles' advice and became too impatient to resist going on the offensive, conducting a disastrous attack on Sicily. Sparta began undermining the Athenian alliance by supporting rebellions, and with reductions in tribute, Athenian naval supremacy was lost.

With a final naval defeat, Athens surrendered. Some Greek city-states argued that the city should be destroyed and all its citizens enslaved, but the Spartans declined. Instead, Athens became a subject state, establishing Sparta as the most powerful city-state among the Greeks. The plague alone had not decided the conflict, but it had permanently tipped the balance of power.

Below: The Athenian plague has captured the imagination of many artists, including Nicolas Poussin (1594–1665).
Opposite: Talisman of an ancient Greek healing god.

7   THE BLACK DEATH

The Black Death, also referred to as the bubonic plague, ranks as the most lethal pandemic in human history, with estimates of the numbers of dead ranging between 100 million and 200 million people in Europe, Asia, and North Africa. Equally remarkable is the fact that this devastation occurred in the short space of only four years, between 1347 and 1351. The changes in population it wrought lasted for two centuries, and its influence is still felt to the present day.

There is evidence that the plague's appearance in 1347 was not its first, and in fact the organism responsible for it has been recovered from European burial remains dating back more than five thousand years. Moreover, recognizable descriptions of the disease are found in ancient texts. The first great pandemic of bubonic plague, the Plague of Justinian, occurred approximately eight hundred years earlier, ravaging the eastern Roman Empire in the years 541 and 542.

TRANSMISSION AND EFFECTS

The microbe behind the Black Death is *Yersinia pestis*, first identified by a Swiss physician at the Pasteur Institute, Alexandre Yersin. The main hosts for the organism are rodents such as marmots and fleas. When a flea bites a rodent, bacteria begin reproducing in its digestive tract. Soon the bacteria become so numerous that they block the flea's digestive tract. When the flea bites a victim, however, it is still capable of transmitting the bacteria by regurgitation.

When infected rodents live in close proximity to humans, the death of the rodent leads the fleas to look to humans as a source of blood. Human cases of plague occur when such a flea moves from a rodent to a human host. Once the bacterium gains access to the bloodstream, it begins to proliferate. The disease is so deadly in part because the organism possesses

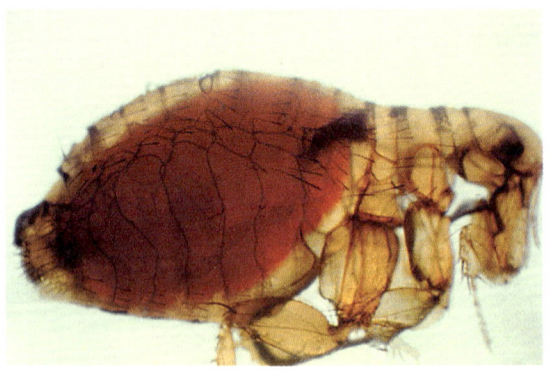

the ability to evade phagocytosis, the process by which white blood cells engulf and destroy bacteria.

The bacterium takes up residence and proliferates in lymph nodes in such locations as the neck, the armpit, and the groin. Lymph nodes enlarge, causing painful swelling in these locations. The infected lymph notes are referred to as buboes, and in some cases they open up to the skin, causing drainage of pus. In addition, patients develop systemic signs of infection, including fever, headaches, and vomiting. Some patients may develop seizures.

Above: The microbe responsible for the Black Death is thought to have been spread by fleas.
Opposite: The *Yersinia pestis* bacterium is named after its discoverer, Alexandre Yersin.

The Black Death

ORIGINS AND SPREAD

It is thought that the Black Death originated in China. In the years preceding its appearance in Europe, it is estimated to have killed between 20 million and 30 million people. Historians hypothesize that it was then spread from east to west along the Silk Road, by which it reached Crimea in 1347. During a siege, traders from the Italian city of Genoa fled the port city of Kaffa and brought the disease to Sicily, then the Italian mainland, from which it spread rapidly.

Within just two years, the disease reached Norway, and in just two more years it had reached all the way to Russia. Ironically, bad weather may have helped disseminate the disease, causing an unusually high number of rat deaths and driving more fleas to alternative hosts, especially human beings. Of course, other factors were involved, including poor nutrition that weakened host immune systems, as well as inadequate hygiene and sanitation, all of which disproportionately afflicted the poor.

It is estimated that perhaps one-third of the population of Europe died during the plague. Large cities, such as London, Paris, and Florence, probably lost between 50 and 80 per cent of their population. More rural, isolated areas tended not to be hit as hard, due to their lower population densities. Of course, not everyone afflicted with the plague died of it, but even today the case fatality rate without treatment is as high as 70 per cent, and it may have been as high as 90 per cent during the Black Death.

People at the time had no idea that bacteria even existed, let alone that they could cause disease, and the role of rats and fleas in spreading the disease was unsuspected. Some authorities argued that astrological or geological forces, such as earthquakes, must be responsible. Others held that the plague was clearly a divine punishment, much like the biblical plagues loosed on the Egyptians. This led many to plead for forgiveness for their sins.

Opposite: Bubonic plague recurred several times throughout history including 1665–6, when it killed a quarter of London's population.
Right, from top: An artist's impression of a flea which carried spores of *Yersinia pestis*; and lead crosses placed on the graves of London plague victims.

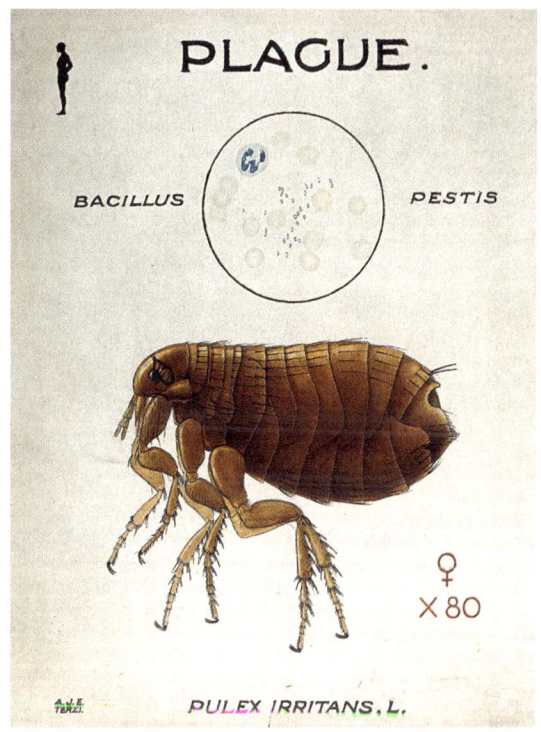

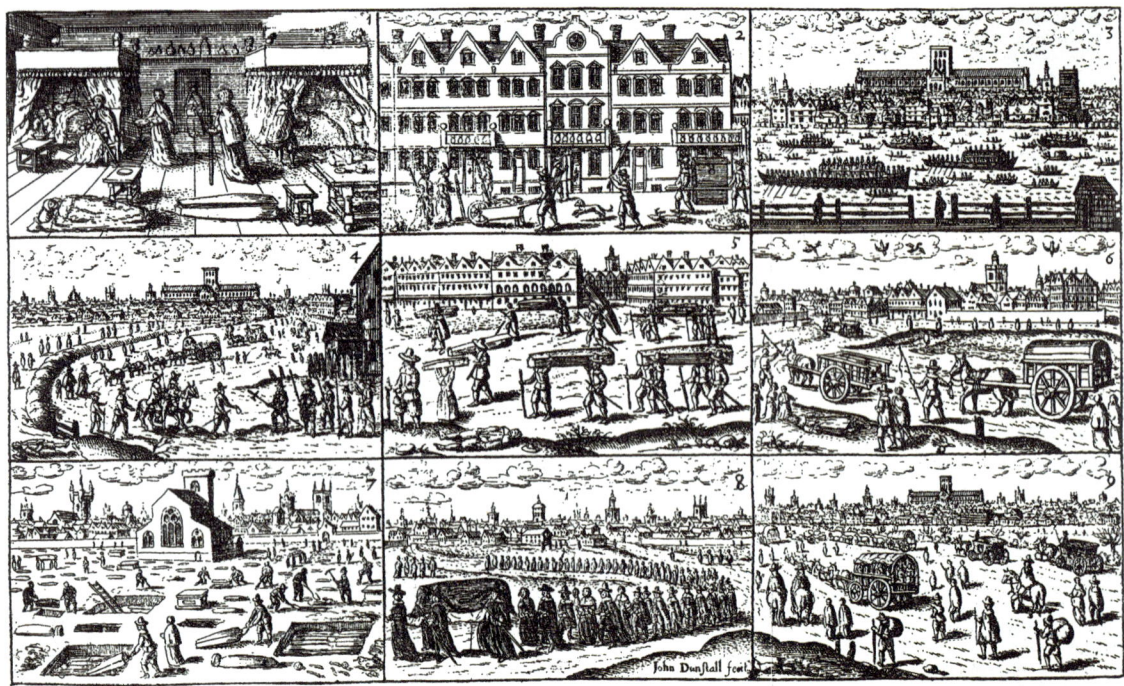

THE WORLD TRANSFORMED

The cultural impact of the Black Death was huge. As landowners expired, their fields were often taken over by peasants. Moreover, the massive reduction in the labour force led to a reordering of society. Fields still needed to be ploughed and crops harvested, so landowners found it necessary to pay wages to attract workers, which helped undermine the traditions of serfdom. Over time, wages rose. In general, peasants began to enjoy greater mobility and enhanced living standards.

The Black Death's influence on art is most manifest in the Danse Macabre, or Dance of Death. Often featuring skeletons frolicking, this genre of art emphasized the universality of death, a fact as timely today as then – after all, the human mortality rate remains stuck at 100 per cent – no one gets out of this life alive. Whether king, pope, titan of industry, famous personage, or poor, anonymous, and powerless child, all eventually meet the same fate.

The Black Death also unleashed waves of violent persecutions, as populations looked for someone to blame for such unprecedented suffering. In the

city of Strasbourg, some blamed Jews for poisoning the wells, leading to a massacre in which perhaps several thousand perished. Others pointed out that the plague seemed to be killing Jews at the same rate as everyone else. Other groups that were persecuted included beggars, foreigners, and people suffering from diseases of the skin.

It should be noted that there is ongoing scholarly debate concerning the cause of the Black Death pandemic. Some historians and scientists assert that while bubonic plague was a key factor, other diseases such as influenza, smallpox, and typhus may have been involved as well. It certainly stands to reason that when an infectious organism wreaks serious damage on a population, such factors as malnutrition and poor sanitation would be exacerbated, leading to other infections taking hold as well.

PLAGUE TODAY

It should also be said that the bubonic plague did not end in 1351. Additional though less severe epidemics and pandemics occurred through much of European history, again claiming millions of lives. Plague has even struck the United States, including an outbreak in San Francisco in the early 1900s. Happily, antibiotic treatment has decreased the mortality rate associated with the disease, which now stands at about 10 per cent.

> **HAPPILY, ANTIBIOTIC TREATMENT HAS DECREASED THE MORTALITY RATE ASSOCIATED WITH THE DISEASE, WHICH NOW STANDS AT ABOUT 10 PER CENT.**

Opposite, from top: A panoramic depiction of the Great Plague of 1665–6 showing mass graves and funeral processions; the plague first appeared in 1346 and killed a third of Europe's population.

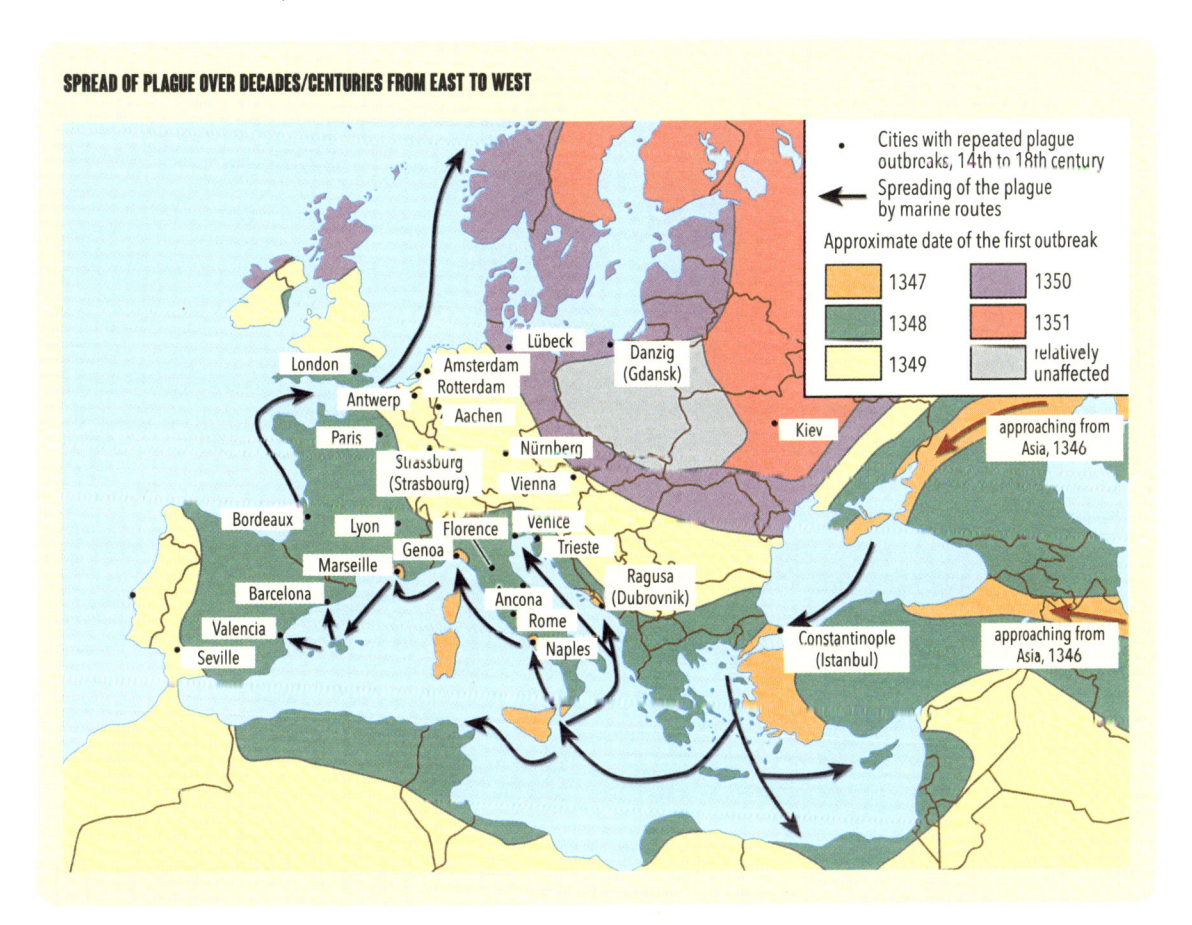

SPREAD OF PLAGUE OVER DECADES/CENTURIES FROM EAST TO WEST

8 BOCCACCIO AND THE BLACK DEATH

Giovanni Boccaccio (1313–75) is author of *The Decameron* and *Concerning Famous Women*, the first work in Western literature focused entirely on women, in which he included over one hundred figures. He was born the illegitimate son of a merchant in the Italian city of Florence. His father wanted him to take up banking but instead he studied law. He soon determined that his true vocation was poetry. The Black Death arrived in Florence in 1348, and he began *The Decameron* the very next year.

The Decameron begins with the arrival of the plague in Florence. Seven young women and three young men decide to escape to a villa in the countryside, where they regale each other with stories, numbering one hundred in total. Some are humorous, while others are deeply sad. Boccaccio's work appears to have influenced Chaucer in the composition of his *Canterbury Tales*. It's title, *Decameron*, is derived from Greek, and alludes to the fact that all the stories are told over a 10-day period.

SYMPTOMS AND SPREAD

Boccaccio describes the plague's symptoms as differing from those described in the East, where it originated.

> It began both in men and women with swellings in the groin or under the armpit. They grew to the size of a small apple or an egg, more or less, and were commonly called tumours. In a short space of time these tumours spread all over the body. Soon after this the symptoms changed and black or purple spots appeared on the arms or thighs or any other part of the body, sometimes a few large ones, sometimes many little ones. These spots were a certain sign of death.

The disease baffled the physicians of the day, who could offer no effective means of prevention or cure.

THE DECAMERON

In addition to its literary merit, *The Decameron* also provides a valuable historical portrait of life at the time, and especially what inhabitants experienced during the Florentine plague. It provides detailed portraits of the affliction's symptoms and signs, the varying manners in which people reacted to it, its effects on social order in general, and the toll it took on burial customs. In many ways, his account of the bubonic plague's effects echoes that of Thucydides' description of the great plague of Athens.

Left: Giovanni Boccaccio wrote some of the most vivid accounts we have of the plague.
Opposite: The use of leeches by doctors was also described in *The Decameron*.

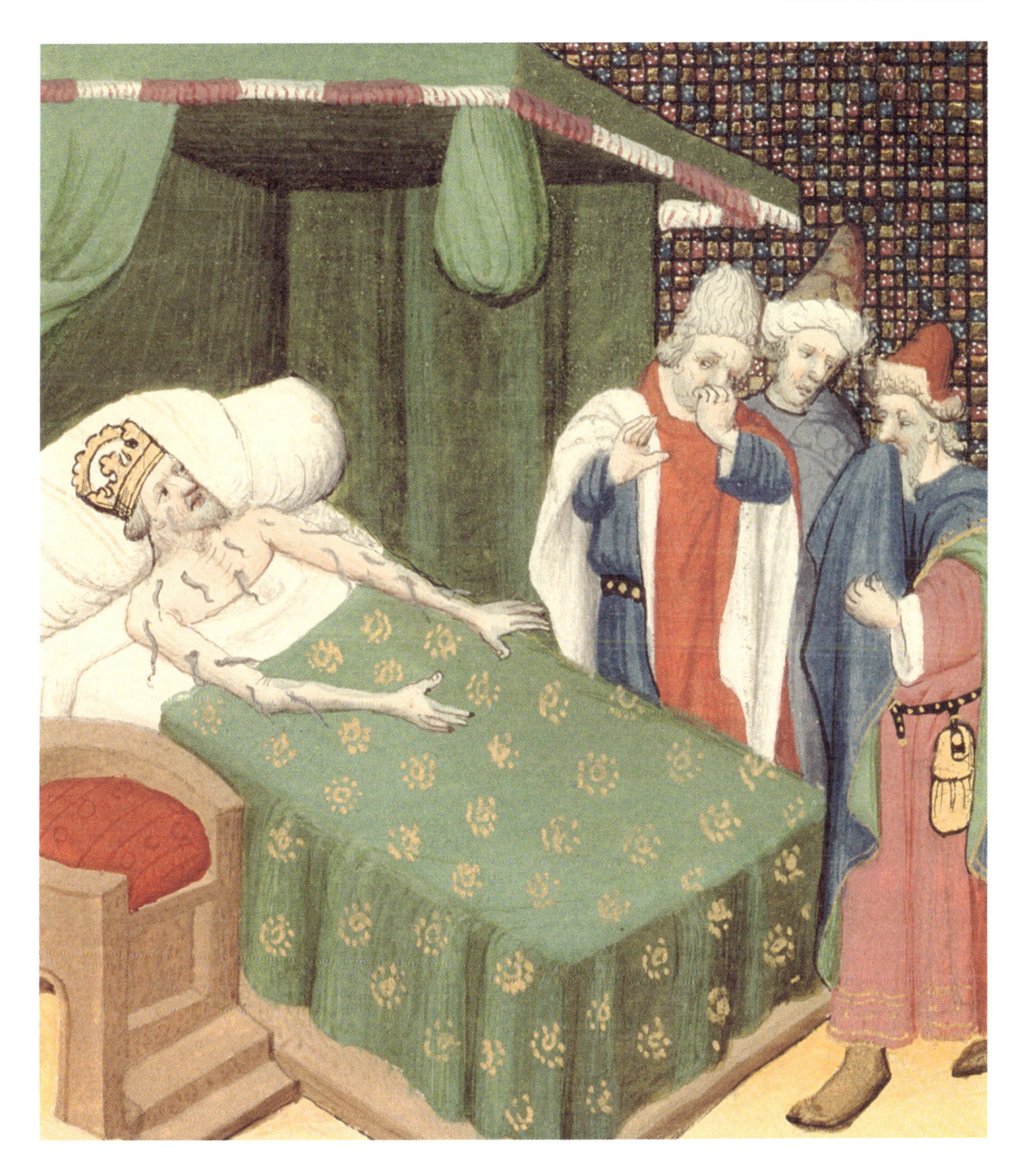

Of particular note was the speed with which the plague could be conveyed from the sick to the healthy – it seemed that anyone who tried to render aid to the sick soon developed the disease.

The sick conveyed it to the healthy who came near them, just as fire catches anything dry or oily near it. And it went even further. Even to speak to or go near the sick brought infection and a common death to the living. Just to touch the clothes or anything else the sick had touched or worn gave the disease to the person who touched it.

FEAR AND PANIC

Not surprisingly, the plague provoked widespread fear and panic among the people of Florence, leading to a general abandonment of the sick. Each person thought first of his or her own safety, leading some to withdraw from contact with others, an approach adopted by the storytellers in *The Decameron*.

> They formed small communities, living entirely separate from everybody else. They shut themselves up in houses where there were no sick, eating the finest food and drinking the best wine very temperately, avoiding all excess, allowing no news or discussion of death and sickness, and passing the time in music and suchlike pleasures.

Others adopted the opposite approach, choosing to remain in the city (though also, as much as possible, avoiding the sick) and enjoying themselves in riotous living, which was made all the easier by the fact that so many people had withdrawn, leaving their property unattended.

> They thought the sure cure for the plague was to drink and be merry, to go about singing and amusing themselves, satisfying every appetite they could, laughing and jesting at what happened. They put their words into practice, spent day and night going from tavern to tavern, drinking immoderately, or going into other people's houses, doing only those things which pleased them. This they could easily do because everyone felt doomed and had abandoned his property, so that most houses became common property and any stranger who went in made use of them as if he had owned them.

As Thucydides described, respect for law and ethics nearly disappeared. Our equivalent of police, judges, and jailors were all either dead or shut up in their homes, hoping to avoid the plague, allowing those who remained to do what they pleased. The impact on social life was devastating – most people scrupulously avoided contact with others, and even relatives and close friends rarely or never visited each other. "Brother abandoned brother, and the uncle his nephew, and the sister her brother, and very often the wife her husband. What is even worse and nearly incredible is that fathers and mothers refused to see and tend their children, as if they had not been theirs."

> # "THE IMPACT ON SOCIAL LIFE WAS DEVASTATING – MOST PEOPLE SCRUPULOUSLY AVOIDED CONTACT WITH OTHERS, AND EVEN RELATIVES AND CLOSE FRIENDS RARELY OR NEVER VISITED EACH OTHER.

BURYING THE DEAD

The high death rate made it at first difficult and soon impossible to observe normal customs surrounding burial. The fact that someone had not been seen or heard from in days meant little, and many "were only known to be dead because the neighbours smelled their decaying bodies". Soon the town was overrun with decaying corpses.

> Survivors were more concerned with getting rid of rotting bodies than moved by charity towards the dead. With the aid of porters, if they could get them, they carried the bodies out of the houses and laid them at the door, where every morning quantities of the dead might be seen … [and] so many corpses were brought every day and almost every hour that there was not enough consecrated ground to give them burial, especially since they wanted to bury each person in the family grave, according to the old custom. Although the cemeteries were full, they were forced to dig huge trenches, where they buried the bodies by the hundreds. Here they stowed them away like bales in the hold of a ship and covered them with a little earth, until the whole trench was full.

In short, vast numbers of afflicted, dying, and dead made it more and more difficult to provide the kind of care people would have expected before the plague arrived. No crowds of women gathered to lament them, "and great numbers passed out of the world without a single person". Instead of shedding tears, even relatives and friends would merely laugh and make merry, for "even women had learned to ignore every other concern and attend only to their own lives".

Having encountered the plague accounts of Thucydides and Boccaccio, readers can only speculate on the toll a deadly pandemic might take on society today. Would people panic? Would respect for and enforcement of laws vanish? Would nurses and physicians remain on duty? And would ordinary people stay at their loved ones' sides to provide care and comfort, or would they abandon them? In plague time, just how fragile might our devotion to the things we hold dear turn out to be?

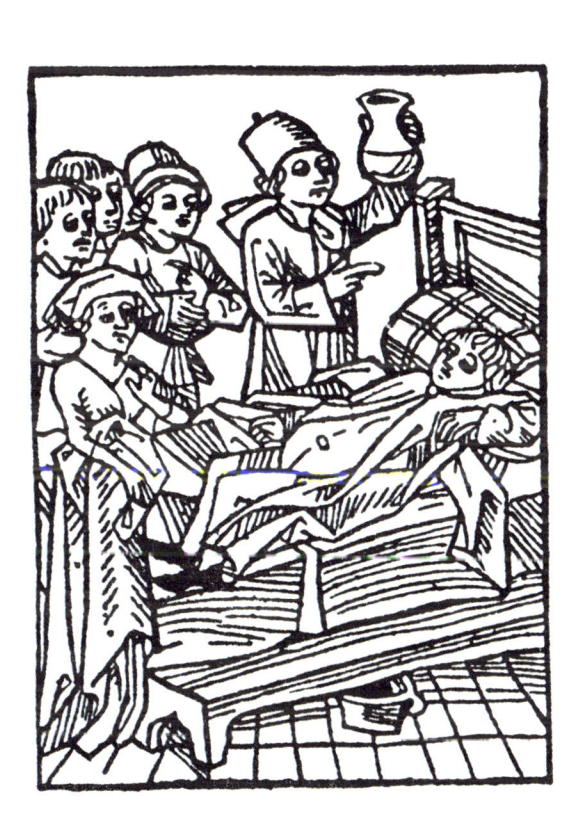

Opposite: The *Decameron* is structured as a series of tales told by seven women and three men.
Above: Another scene from *The Decameron* related to the Black Death.
Following pages: A depiction of the Plague of Florence in 1348, which Boccaccio lived through.

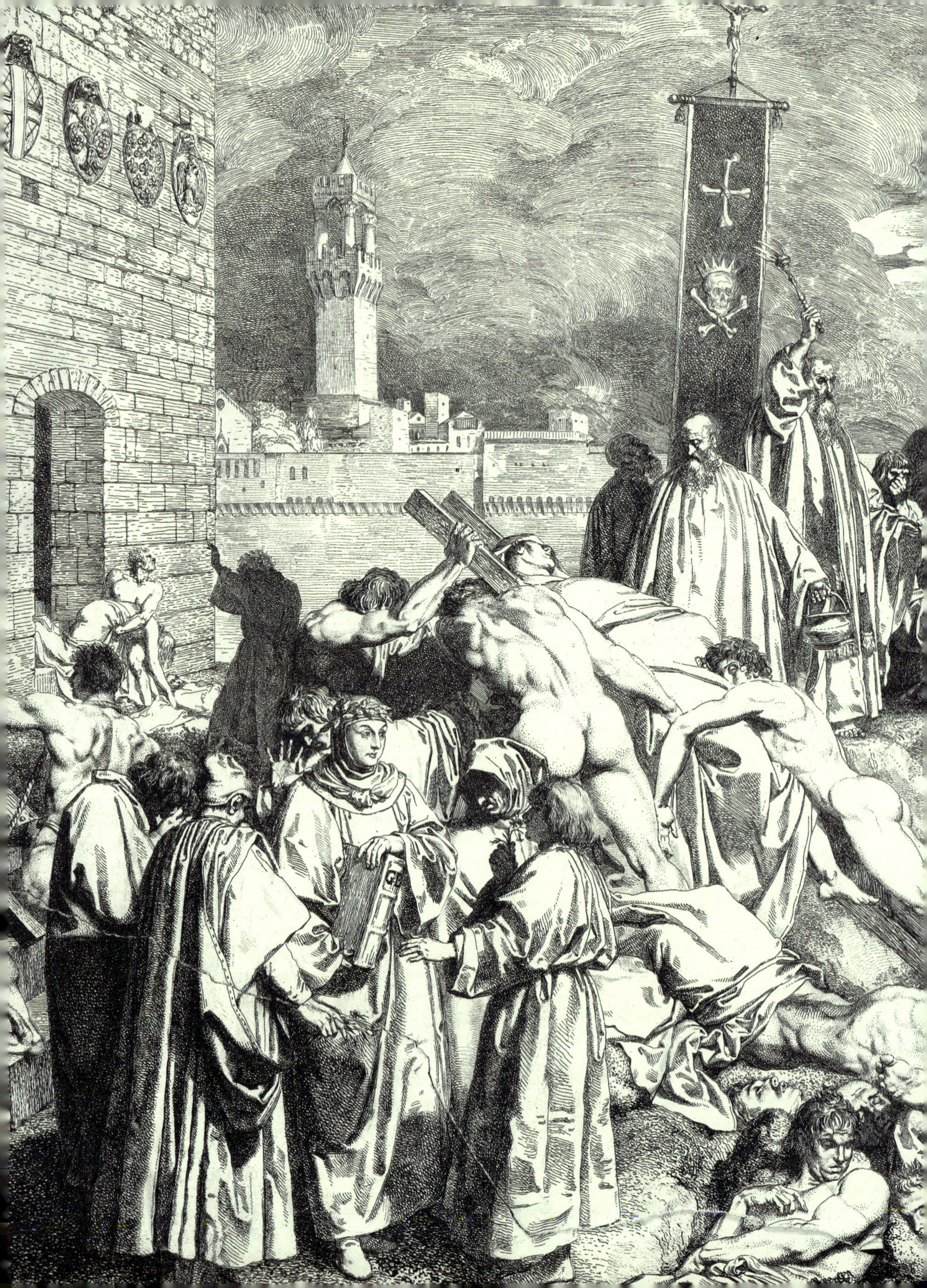

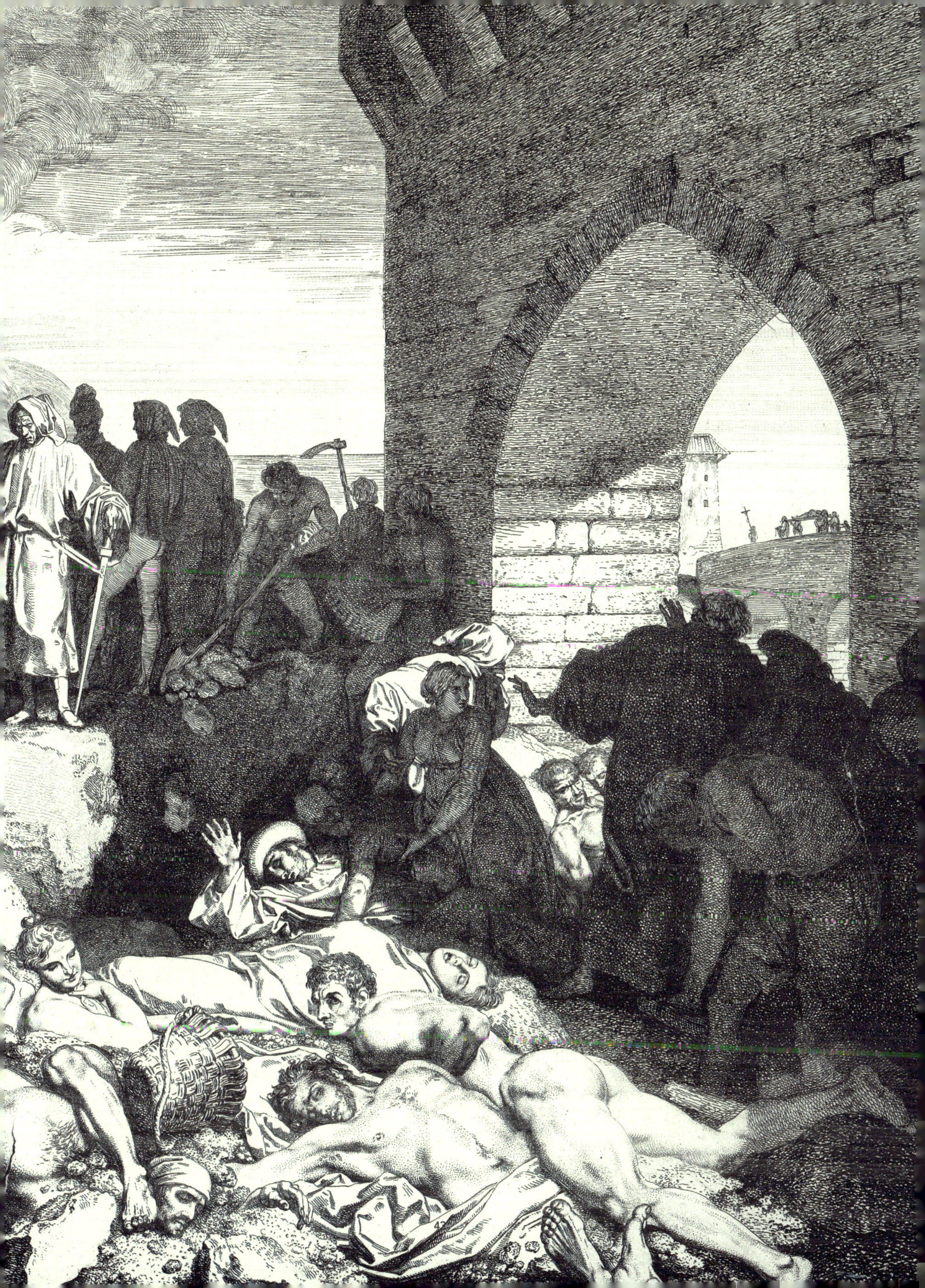

9 SPAIN'S CONQUEST OF THE AZTECS

Five hundred years ago, in February of 1519, the Spaniard Hernán Cortés set sail from Cuba to explore and colonize Aztec civilization in the Mexican interior. Within just two years, Aztec ruler Montezuma was dead, the capital city of Tenochtitlan was captured, and Cortés had claimed the Aztec empire for Spain. Spanish weaponry and tactics played a role, but most of the destruction was wrought by epidemics of European diseases.

After helping conquer Cuba for the Spanish, Cortés was commissioned to lead an expedition to the mainland. When his small fleet landed, he ordered his ships scuttled, eliminating any possibility of retreat and conveying the depth of his resolve.

Cortés with his 500 men then headed into the Mexican interior. This region was home to the Aztec civilization, an empire of an estimated 16 million people at the time. Through a system of conquest and tribute, the Aztecs had established the great island city of Tenochtitlan in Lake Texcoco that ruled over an area of about eighty thousand square miles.

Discovering widespread resentment towards the capital city and its ruler, Cortés formed alliances with many locals. Though vastly outnumbered, he and a small force marched on Tenochtitlan, where Montezuma received them with honour. In return, Cortés took Montezuma prisoner.

It took Cortés two years, but he finally conquered the Aztec capital in August 1521. His allies in this fight were the European germs he and his men unwittingly brought with them.

Although Cortés was a skilled leader, he and his force of perhaps a thousand Spaniards and indigenous allies would not have been able to overcome a city of 200,000 without help. He got it in the form of a smallpox epidemic that gradually spread inwards from the coast of Mexico and decimated the densely populated city of Tenochtitlan in 1520, reducing its population by 40 per cent in a single year.

HOW SMALLPOX CHANGED THE WORLD

Smallpox is caused by an inhaled virus, which causes fever, vomiting, and a rash, soon covering the body with fluid-filled blisters. These turn into scabs that leave scars. Fatal in approximately one-third of cases, another third of those afflicted with the disease typically develop blindness.

Smallpox existed in ancient times in Egyptian, Indian, and Chinese cultures. It remained endemic in human populations for millennia, coming to Europe during the eleventh century's Crusades. When Europeans began to explore and colonize other parts of the world, smallpox travelled with them.

Above: Montezuma, the last Aztec emperor, is here depicted in an eighteenth-century painting.
Opposite: A nineteenth-century lithograph featuring the first meeting between Montezuma and Cortes in 1519.

The native people of the Americas, including the Aztecs, were especially vulnerable to smallpox, because they had never been exposed to the virus and thus possessed no natural immunity. No effective antiviral therapies were available.

Recalling the epidemic, one victim reported: "The plague lasted for 70 days, striking everywhere in the city and killing a vast number of our people. Sores erupted on our faces, our breasts, our bellies; we were covered with agonizing sores from head to foot." A Franciscan monk who accompanied Cortés provided this description:

As the Indians did not know the remedy of the disease, they died in heaps, like bedbugs. In many places it happened that everyone in a house died, and as it was impossible to bury the great number of dead, they pulled down the houses over them, so that their homes became their tombs.

Smallpox took its toll on the Aztecs in several ways. First, it killed many of its victims outright, particularly infants and young children. Many adults were incapacitated by the disease – because they were either sick themselves, caring for sick relatives and neighbours, or simply lost the will to resist the Spaniards as they saw disease ravage those around them. Finally, people could no longer tend to their crops, which led to widespread famine, further weakening the immune systems of survivors.

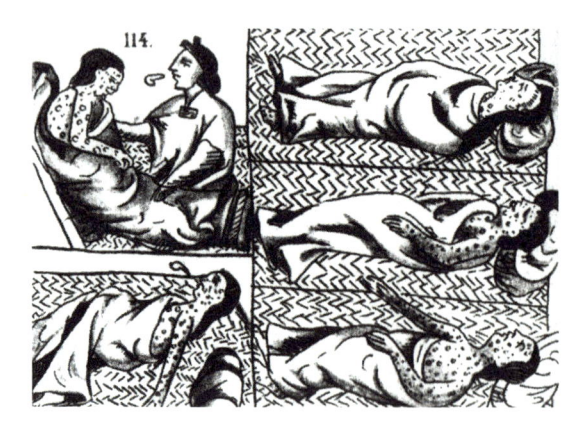

MORE VICTIMS

Of course, the Aztecs were not the only indigenous people to suffer from the introduction of European diseases. In addition to North America's Native American populations, the Mayan and Incan civilizations were also nearly wiped out by smallpox. And other European diseases, such as measles and mumps, also took substantial tolls – altogether reducing some indigenous populations in the New World by 90 per cent or more. Recent investigations have suggested that various infectious agents, such as Salmonella – known for causing outbreaks today among pet owners – may have caused additional epidemics.

The ability of smallpox to incapacitate and decimate populations made it an attractive agent for biological warfare. In the eighteenth century, the British tried to infect Native American populations. One commander wrote, "We gave them two blankets and a handkerchief out of the smallpox hospital. I hope it will have the desired effect." During the Second World War, British, American, Japanese, and Soviet teams all investigated the possibility of producing a smallpox biological weapon.

Happily, worldwide vaccination efforts have been successful. The final case of smallpox occurred in 1978, when a photographer died of the disease, prompting the scientist whose research she was covering to take his own life.

Many great encounters in world history, including Cortés's clash with the Aztec empire, had less to do with weaponry, tactics, and strategy than the ravages of disease. Nations that suppose they can secure themselves strictly through investments in military spending should study history – time and time again, the course of events has been definitively altered by disease outbreaks. Microbes too small to be seen by the naked eye can render ineffectual even the mightiest machinery of war.

Spain's Conquest of the Aztecs

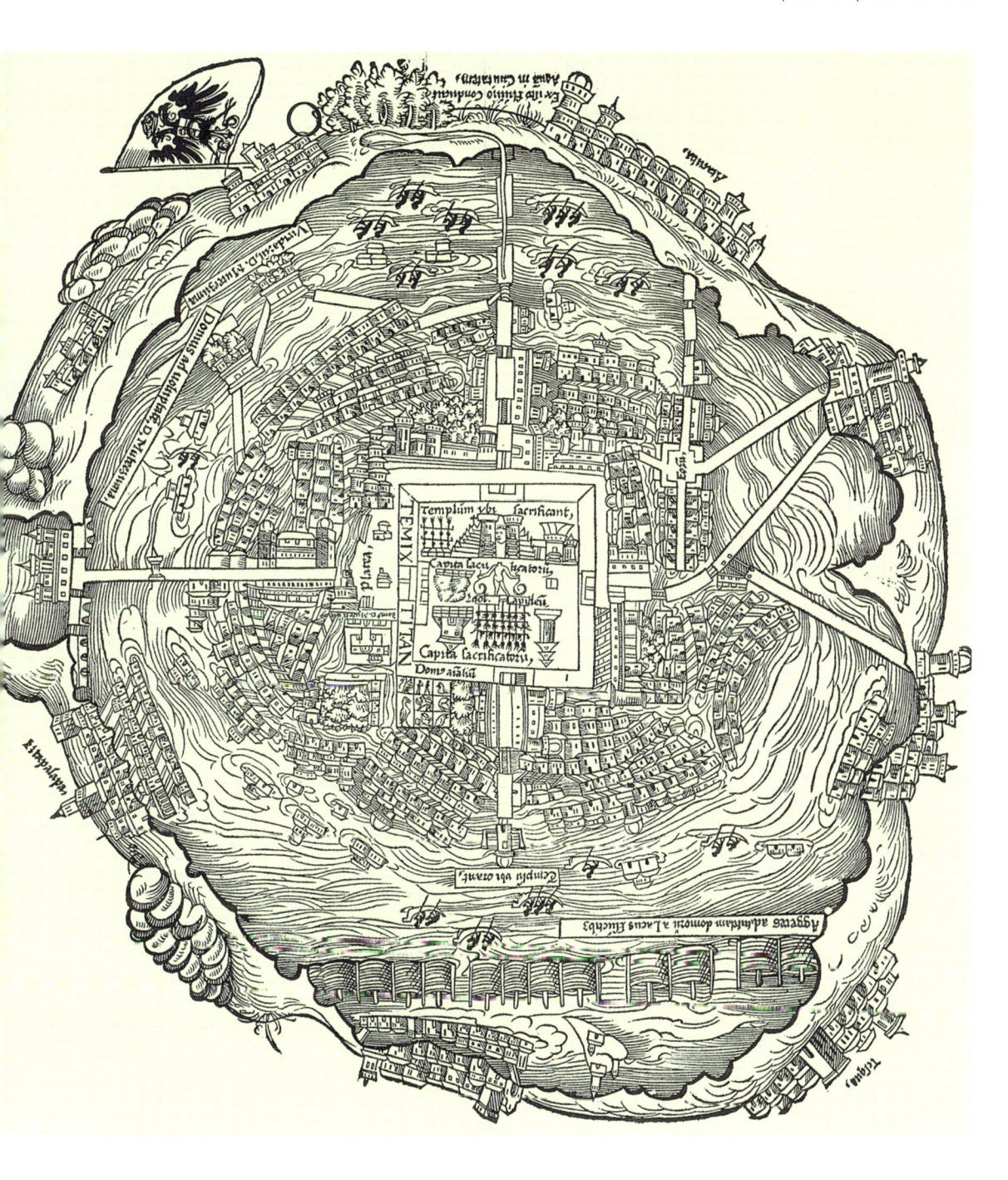

Opposite: The Aztecs quickly succumbed to European smallpox, to which they had no immunity.
Above: A map of the Mexican capital Tenochtitlan.

PATTERNS OF DEATH FROM SMALLPOX
decade in which smallpox ceased to be endemic by country

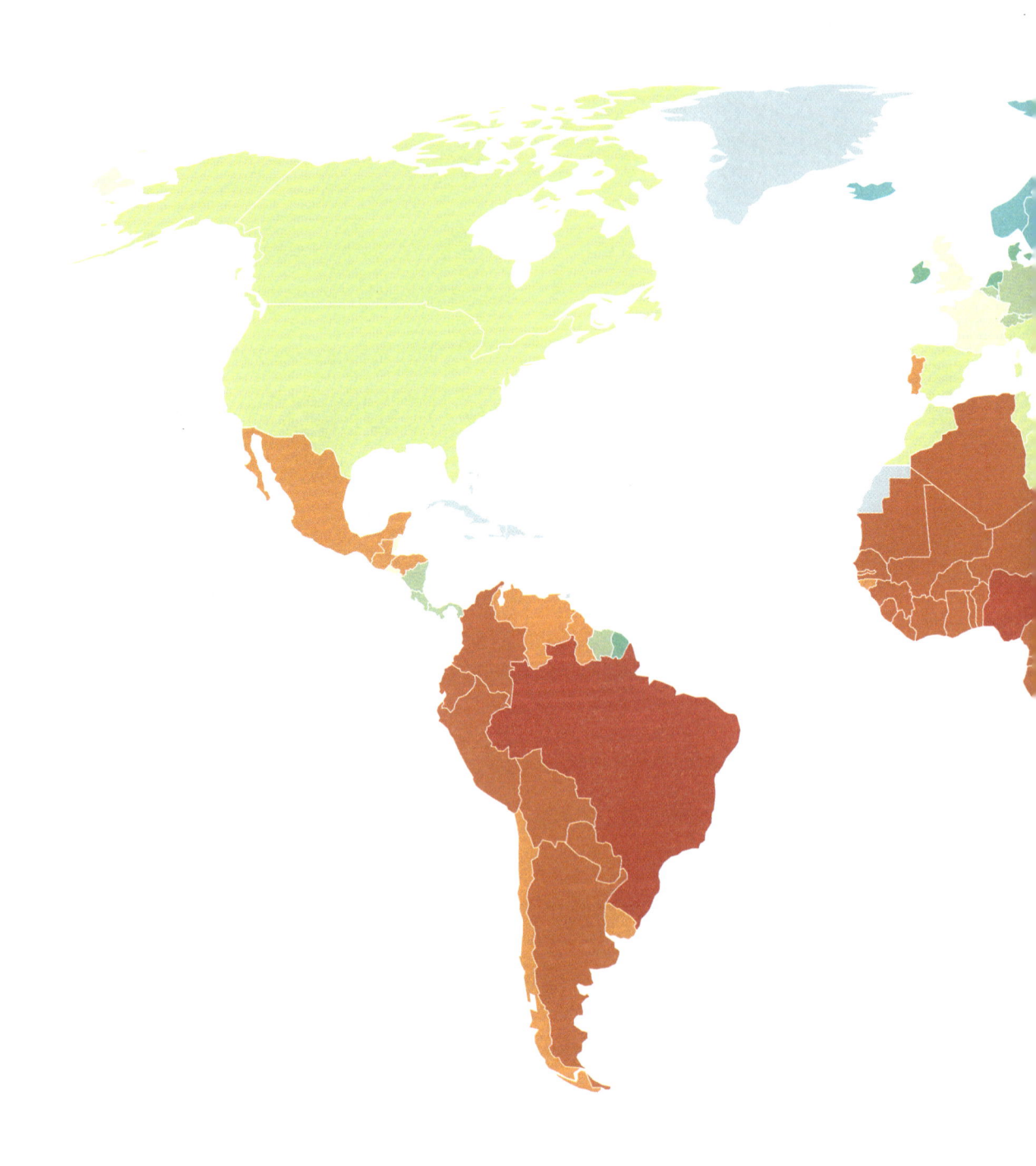

Spain's Conquest of the Aztecs

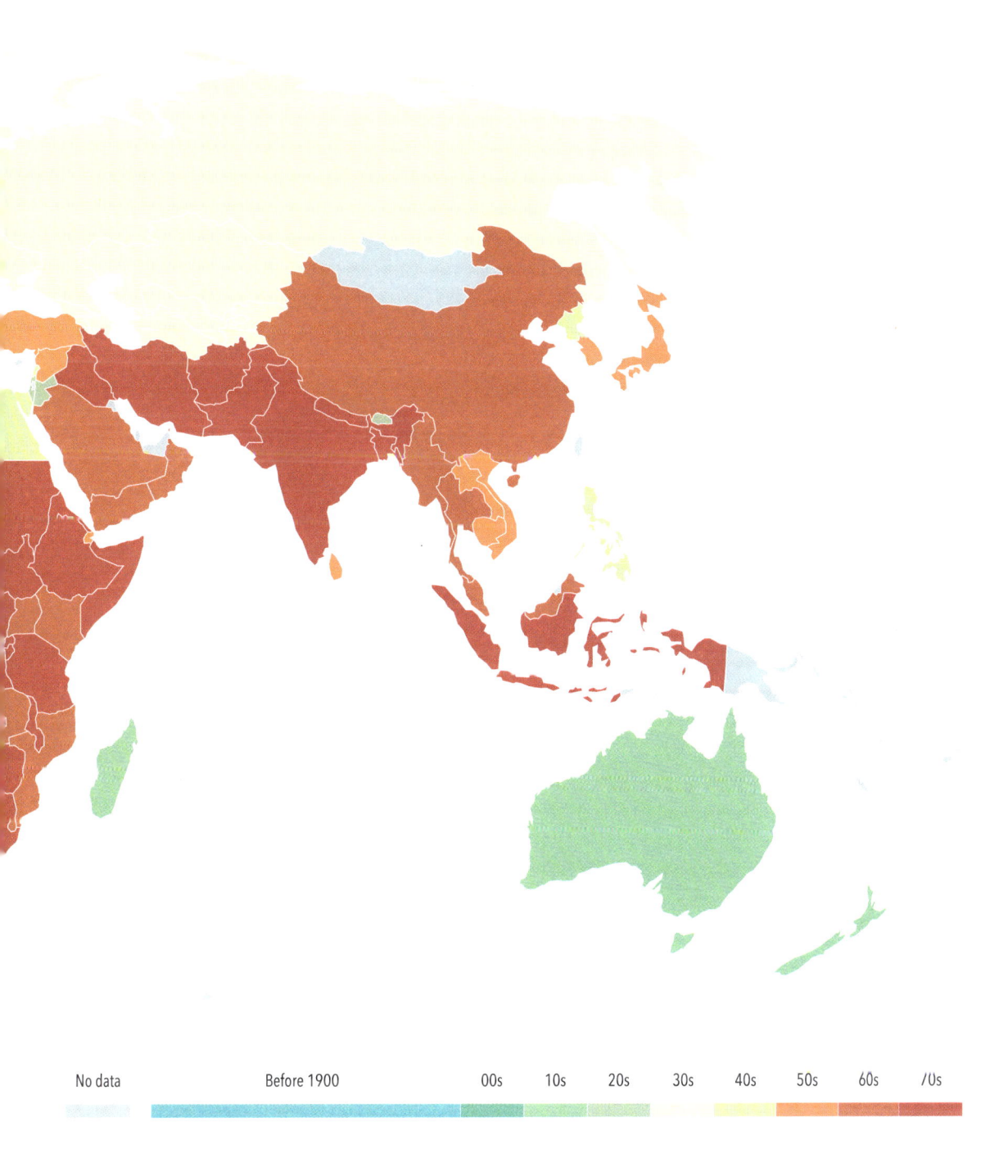

| No data | Before 1900 | 00s | 10s | 20s | 30s | 40s | 50s | 60s | 70s |

10 SEEING MICROBES FOR THE FIRST TIME: LEEUWENHOEK

The word microbe comes from two Greek roots, meaning "small" and "life". For most of human history, it was practically impossible to trace infectious diseases to microbes, because no one could have known that there are life forms too small to be seen by the naked eye. Discolouration and films on water, the decay of foodstuffs and dead organisms, and foul odours associated with them all suggested that putrid forces might be at work, but their source remained a mystery.

EARLY MICROSCOPES

Hence the story of the invention and development of microscopy is a vital chapter in our understanding of infectious disease – no microscope, no microbiology. Unfortunately, no one knows who invented the microscope. The ancient Greeks discussed the power of water droplets to magnify, and medieval eyeglass makers were aware of the magnifying power of lenses. A period of rapid innovation among lens makers arose in Holland and Italy during the early 1600s.

In about 1610, Galileo (1564–1642) seems to have realized that his telescope could be trained on terrestrial objects, using it to produce magnified views of insects. A decade or so later, he appears to have been using a compound microscope, which used two sets of lenses to obtain a higher degree of magnification. By the mid-1620s, colleagues with whom he shared this innovation coined the term microscope, from the Greek for "to look at small things".

THE FATHER OF MICROBIOLOGY

From a biological point of view, perhaps the most important early microscopist was Antonie van Leeuwenhoek, a Dutch draper who is sometimes referred to as the "father of microbiology". Born in Delft in 1632, a contemporary of the painter Vermeer, Leeuwenhoek developed an interest in using lenses to better visualize the thread he was using in his business. He developed a technique for creating tiny, high-quality glass spheres, superior to the lenses of the day.

Above: Galileo used lenses to see distant objects.
Right: Leeuwenhoek used lenses to see very small objects.
Opposite: Some remarkable images from Robert Hooke's *Micrographia*.

> **IT WAS LEEUWENHOEK WHO ORIGINALLY REPORTED THE EXISTENCE OF SINGLE-CELLED ORGANISMS, A CLAIM THAT WAS INITIALLY MET WITH INCREDULITY BY THE [ROYAL SOCIETY'S] MEMBERS.**

Leeuwenhoek made hundreds of lenses and created several dozen single-lens microscopes, of which nine are still in existence. The best of these microscopes can magnify objects up to 300 times, and some suspect that some of his other devices may have been even more powerful. Leeuwenhoek himself kept some aspects of his microscope construction technique secret. Remarkably, he and other microscopists of the age relied on sunlight to illuminate their specimens.

Regarding himself as a craftsman and businessman with no scientific training, Leeuwenhoek initially expressed reservations about sharing his early work in microscopy. But when his work was brought to the attention of the Royal Society in London, he began describing his findings in letters. It was Leeuwenhoek who originally reported the existence of single-celled organisms, a claim that was initially met with incredulity by the society's members.

Although the term "cell" was coined by a British contemporary, Robert Hooke (1635–1703), Leeuwenhoek is generally acknowledged to be the first person to visualize numerous biological entities. These include spermatozoa and, most important for present purposes, bacteria, which he called "animalcules", meaning little animals. He may have been the first person to stain his specimens to improve visualization, using the spice saffron.

Here is Leeuwenhoek describing his discovery in a letter in 1676:

> I now saw very distinctly that these were little eels or worms … Lying huddled together and wriggling, just as if you saw with your naked eye a whole tubful of very little eels and water, the eels moving about in swarms; and the whole water seemed to be alive with the multitudinous animalcules. For me this was among all the marvels that I have discovered in nature the most marvellous of all, and I must say that, for my part, no more pleasant sight has yet met my eye than this of so many thousands of living creatures in one small drop of water, all huddling and moving, but each creature having its own motion.

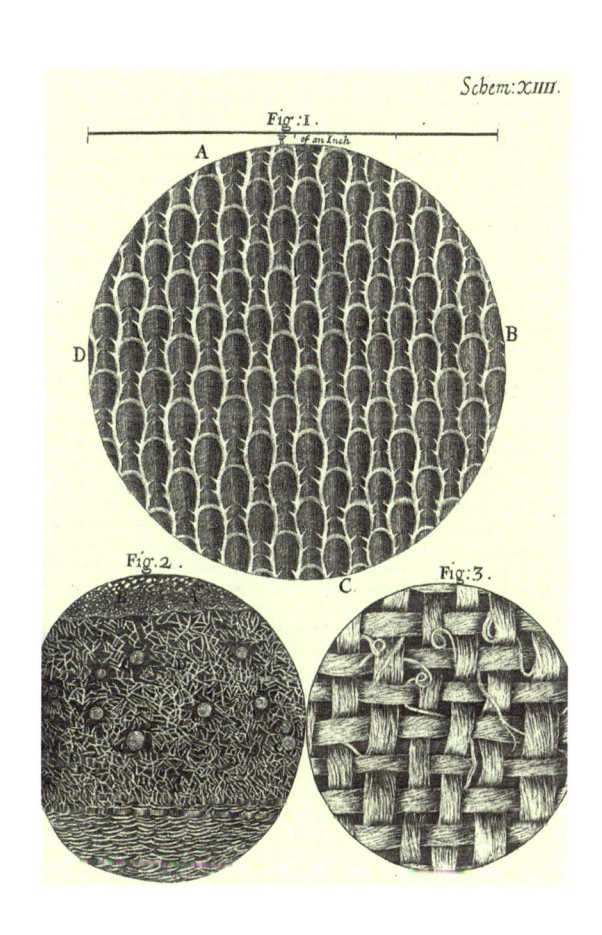

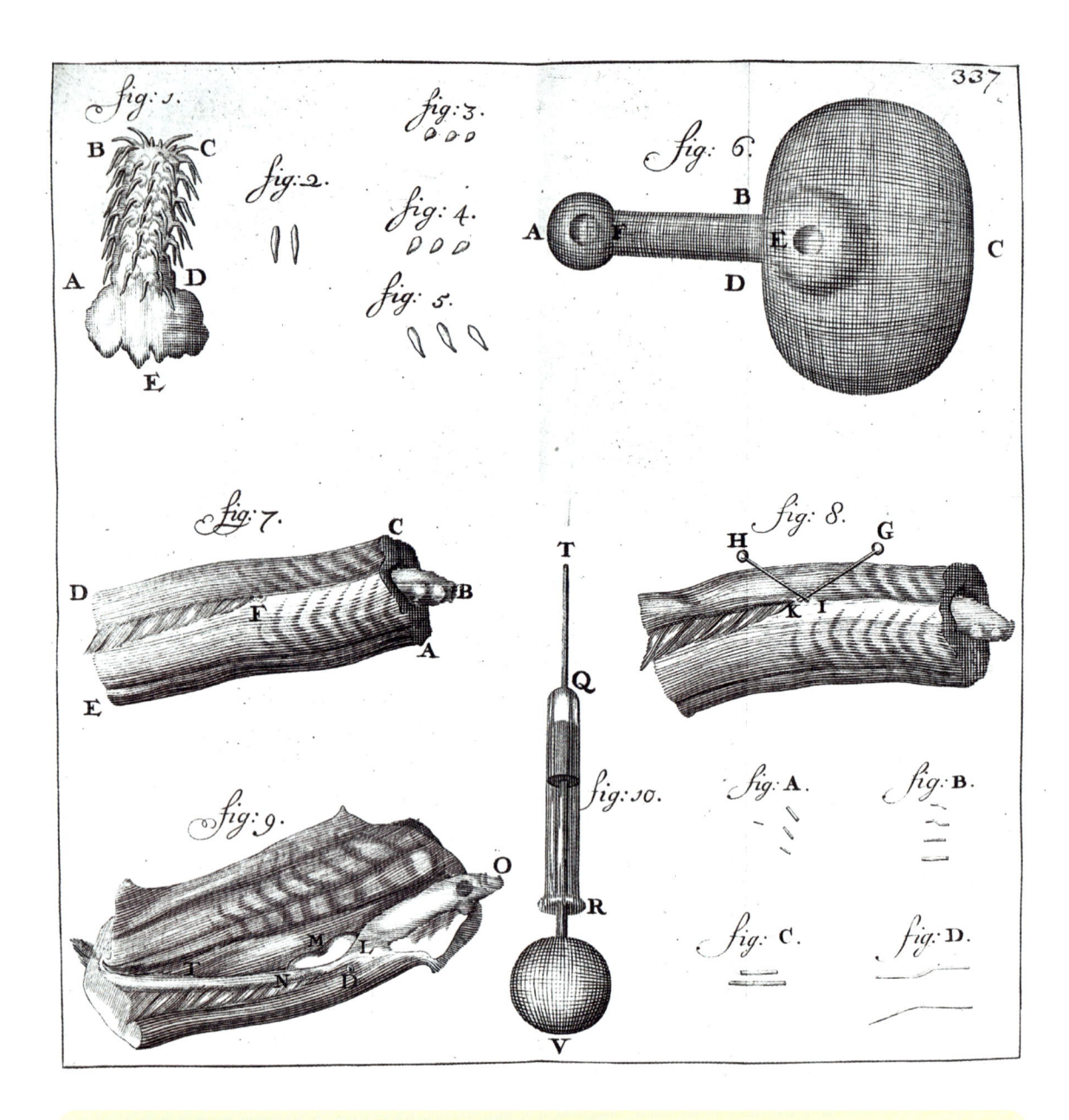

A LIFE WELL LIVED

Leeuwenhoek lived to the age of 90 years. Reflecting on what he had accomplished over the course of his lifetime and what had driven him to do so, he wrote:

> My work, which I have done for a long time, was not pursued in order to gain the praise that I now enjoy, but chiefly from a craving after knowledge, which I notice resides in me more than in most other men. And therefore, whenever I found out anything remarkable, I have thought it my duty to put down my discovery on paper, so that all ingenious people might know of it.

A MAJOR LEGACY

Despite Leeuwenhoek's acute awareness of his own lack of scientific qualifications, he later joined such notable natural philosophers as Hooke, Robert Boyle (1627–91, the first modern chemist), and Christopher Wren (1632–1723, acclaimed architect who designed St Paul's Cathedral in London) as a full member of the Royal Society, even though he never attended a meeting. Today, a cancer hospital in Amsterdam bears his name, and he is commemorated with both a medal and a lecture.

ELECTRON MICROSCOPES

Since Leeuwenhoek, microscopy has been developed and refined. Light microscopy was eventually taken to its limit, which is based on the wavelength of visible light (400–700 nm). This permits it to magnify specimens between up to 500 to 1,500 times. In the twentieth century, electron microscopes were introduced. Because electrons have a wavelength of approximately 1 nm, they can achieve a much higher degree of magnification, ranging between 16,000 and 1 million times.

Though much more powerful than light microscopes, electron microscopes do have drawbacks. For example, unlike the light microscopes used by Leeuwenhoek, they can only visualize dead organisms and tissues. But electron microscopes can create exquisite images of much smaller structures, such as the tiny organelles within cells. Of special note in microbiology, electron microscopes also introduced the ability to visualize most viruses.

Today, microscopes continue to play a vital role in science, and especially in microbiology. To this day, we say that we know something when we are able to see it, and the invention and refinement of the microscope has made it possible to see and know a previously unknown dimension that is positively teeming with life. Once we learned that microbes exist, it became possible to investigate their role in health and disease.

Opposite: Some of Leeuwenhoek's observations of bacteria.
Right, from top: Robert Hooke's *Micrographia* is perhaps the most celebrated work of microscopy; one of the apparatuses used by Leeuwenhoek.

11 SMALLPOX INOCULATION AND THE AMERICAN FOUNDING

WASHINGTON AND SMALLPOX

The experience of George Washington (1732–99), commander of American troops during the War of Independence with England and first president of the United States, offers important insights into the history of inoculation and its role in world history. Washington was born into a wealthy family, with five siblings and three half-siblings from his father's first marriage. One of the latter was his brother Lawrence, 12 years his senior and heir to the estate now known as Mount Vernon.

Lawrence was afflicted with consumption (tuberculosis) and had been advised by physicians that a move to a warmer climate represented his best hope for survival. The 19-year-old George, who had always benefited from his older half-brother's mentorship, agreed to accompany him to Barbados in the Caribbean – his only trip outside the continent. While there, Washington developed a high fever and severe headache. He had contracted smallpox.

Washington's case was considered mild at the time, but it left him housebound for 25 days. Just a week after his recovery, he returned to Virginia. Lawrence voyaged to Bermuda, hoping for better luck there, but his illness persisted, and he died within a year. His wife and young daughter were the heirs to his estate, but both survived only two more years, leaving Mount Vernon to Washington. Although his experience in Barbados was not a happy one, it left Washington with a great gift: immunity from smallpox.

Above: George Washington gained early immunity from smallpox.
Right: Washington's half-brother, Lawrence, also suffered from TB.
Opposite: A cartoon showing Death scattering pestilence among the people.

> **ALTHOUGH HIS EXPERIENCE IN BARBADOS [– WHERE HE CONTRACTED SMALLPOX –] WAS NOT A HAPPY ONE, IT LEFT WASHINGTON WITH A GREAT GIFT: IMMUNITY FROM [THE DISEASE].**

THE QUARANTINE QUESTION.
DEATH, RISING FROM THE IRON SCOW, AND SCATTERING PESTILENCE AMONG THE PEOPLE.

THE WAR OF INDEPENDENCE

That gift proved priceless some years later, when Washington took command of the troops during the War of Independence. Because the colonies were so rural, with most people living widely dispersed, epidemics of smallpox were relatively unknown. Only cities such as Boston and Philadelphia saw large numbers of cases. But as American troops began to amass in large numbers, crowding made conditions favourable for outbreaks. European troops, having already contracted the disease, were immune.

In the fall of 1775, an epidemic of smallpox broke out in Boston. Some suspected that the British might be fostering the epidemic, by sending infected people into the cities. To prevent the spread of disease, Washington, who had assumed command of the Continental forces in June, prohibited anyone fleeing British occupation from entering the American camp.

Washington wrote to a friend, "If we escape the smallpox in this camp and the country round about it, it will be a miracle."

Once the British left Boston the next year, he sent troops who had survived smallpox – and were therefore immune – to occupy the city. The next year, epidemics arose in Boston and Philadelphia. Washington knew the disease could exert a greater effect on the outcome of the war than any battle plan or tactics. He was a believer in inoculation and had made it a regular practice among his own slaves back in Virginia.

INOCULATION

Inoculation was performed by scraping the pustules of a patient with a mild case of smallpox, then passing it on a thread underneath the skin of a person who had never suffered the disease. The usual result was a mild case of the disease. This procedure was known as variolation, variola being another name for smallpox. Edward Jenner's technique of cowpox inoculation was referred to as vaccination. Later, vaccination would become synonymous with immunization.

Despite his belief in inoculation, in May of 1776 Washington gave orders that no troops were to be inoculated. He knew that fighting would resume in the summer, and he could not afford to have large numbers of his troops incapacitated. But he also knew that eventually he would need to act to protect his troops against the scourge, an outbreak of which could decimate his army and doom the American cause.

Having seen how the disease could spread through an encampment like wildfire, however, Washington did adopt aggressive quarantine measures. Along with those suffering from other diseases such as malaria and diphtheria, afflicted troops were removed from camps and housed at a distance in quarantine hospitals. Perhaps unwittingly, Washington was helping to lay the groundwork for public health initiatives in the New World.

Having for some time assured the Continental Congress that he was exercising his "utmost vigilance against this most dangerous enemy", in January of 1777 Washington ordered physicians to inoculate all the troops and new recruits who had not been previously infected. He wrote: "Necessity not only authorizes but seems to require the measure, for should the disorder infect the army in the natural way and rage with its usual virulence, we should have more to dread from it than the sword of the enemy."

While the story of such wars is usually told in terms of military strategy and battle outcomes, it is quite possible that this decision played as important a role in the eventual victory of the revolutionary forces as any more conventionally military one Washington achieved. After all, approximately 90 per cent of deaths among the Continental forces were not the result of injuries sustained in battle but due to infectious diseases.

This marked the first large-scale military inoculation programme in world history. Washington understood the risk he was taking and took great pains to ensure that not only the programme itself but all communication concerning it was kept highly secret. The inoculation programme would not only secure the immunity of active troops but also help to allay the fears of new recruits, who had heard rumours of higher rates of disease among those already in service.

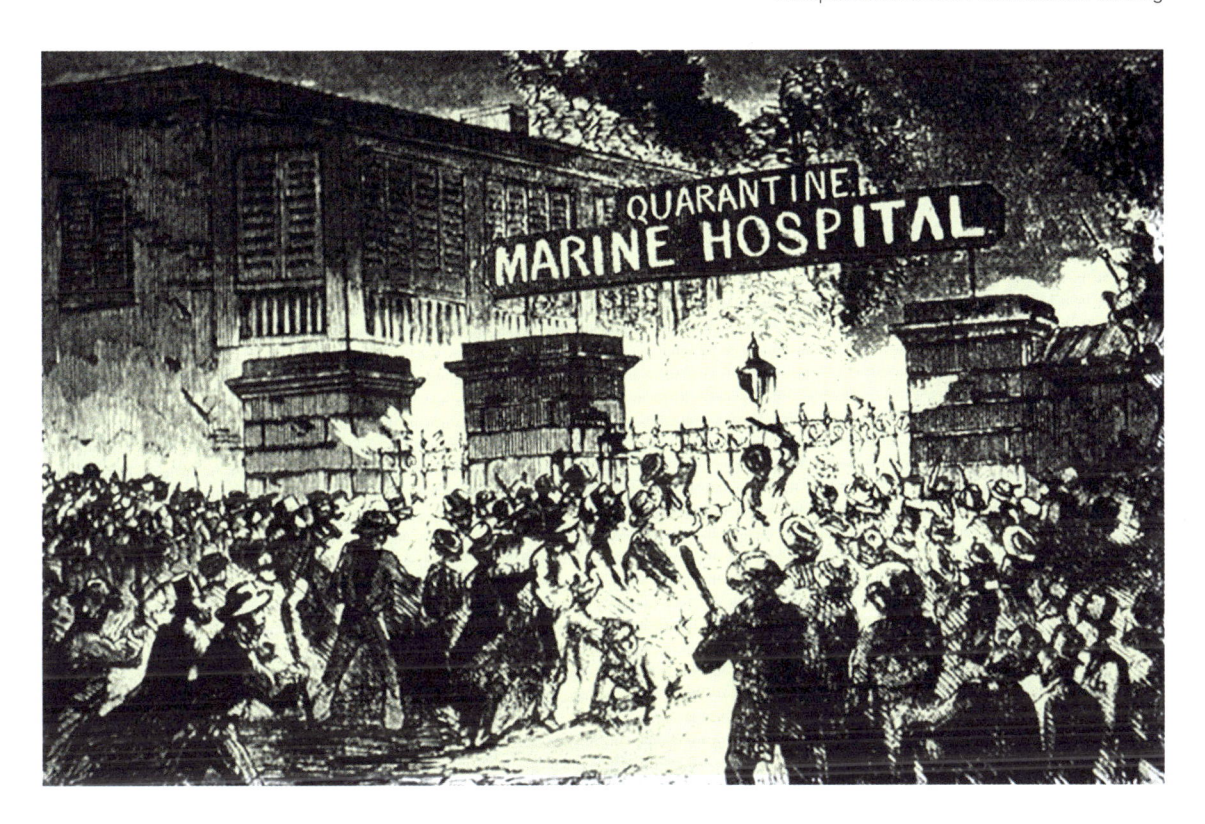

A smallpox epidemic raged in the colonies between 1775 and 1782, quickly spreading throughout the continent of North America. Many Native Americans and slaves seeking release from bondage served in the British forces, and smallpox death rates among them were high. Native American tribes, which lacked prior exposure to the disease, were especially hard hit, reducing the population in some areas by as much as a third.

Other American leaders had also faced the inoculation decision. Washington's later successor as US president, John Adams, chose to undergo the procedure relatively early, in 1764. He described the procedure to his wife, Abigail:

> [The physicians] took their lancets and with their points divided the skin about a quarter of an inch and just suffering the blood to appear, buried a thread (infected) about a quarter of an inch along the channel … Do not conclude from anything I have written that I think inoculation a light matter – a long and total abstinence from everything in nature that has any taste; two long heavy vomits … and three weeks of close confinement to a house are, according to my estimation, no small matters.

Other American founders made different choices. For example, in 1736 Benjamin Franklin, perhaps the most accomplished of all the founders, chose to forgo smallpox inoculation for his son, Francis, a decision he regretted for the rest of his life, writing:

> In 1736 I lost one of my sons, a fine boy of four years old, by the smallpox taken in the common way. I long regretted bitterly and still regret that I had not given it to him by inoculation. This I mention for the sake of the parents who omit that operation, on the supposition that they should never forgive themselves if a child died under it; my example showing that the regret may be the same either way, and that, therefore, the safer should be chosen.

Opposite, from top: Washington at the Continental Army's winter encampment in New Jersey; Ben Franklin refused the smallpox inoculation for his son.
Above: From 1858, New York's Quarantine hospital was the first line of defence for contagion among immigrants.

12

THE PHYSICIAN AS HERO IN PLAGUE TIME: DR BENJAMIN RUSH

Benjamin Rush (1746–1813) stands out as perhaps the greatest physician-public servant in US history. A prodigy who still ranks as the youngest graduate in the history of Princeton University, author of the first American textbook of chemistry, one of the youngest signers of the Declaration of Independence, Treasurer of the US Mint, "father of American psychiatry", founder of Dickinson College and namesake of Rush Medical College in Chicago, and reconciler of John Adams and Thomas Jefferson, Rush is perhaps most deserving of attention today for his response to one of the deadliest epidemics in US history, the yellow fever outbreak that decimated the nation's capital in 1793.

DR BENJAMIN RUSH: BACKGROUND AND LEGACY

Rush was born in a township of Philadelphia, the fourth of seven children. His father died when he was five. Rush graduated from what is now Princeton at age 14, then studied medicine at the University of Edinburgh, becoming fluent in multiple languages during his tour of Europe. When he returned in 1769, he set up his medical practice and became a chemistry professor at what is now the University of Pennsylvania, publishing the first American textbook in the field. As an elected delegate to the Continental Congress, Rush was a strong proponent of American independence and encouraged Thomas Paine to write the widely influential pamphlet, "Common Sense". He signed the Declaration of Independence.

During the War of Independence, Rush served as surgeon-general in the Continental Army, promoting a variety of reforms to improve the health of soldiers. After the war, Rush served on the staff at Pennsylvania Hospital and resumed his duties as a chemistry professor at the University of Pennsylvania. When Thomas Jefferson was commissioning Captain Meriwether Lewis and Lieutenant William Clark for their epic expedition across the country, he sent them to Rush to obtain necessary medical training and supplies. Rush was a staunch abolitionist, arguing that blacks were in no way naturally inferior to whites, and he campaigned against capital punishment and promoted the education of women.

One of the most prominent mental health reformers in US history, in 1812 Rush published his *Medical Inquiries and Observations Upon the Diseases of the Mind*. He deplored the conditions under which many psychiatric patients were kept and lobbied for

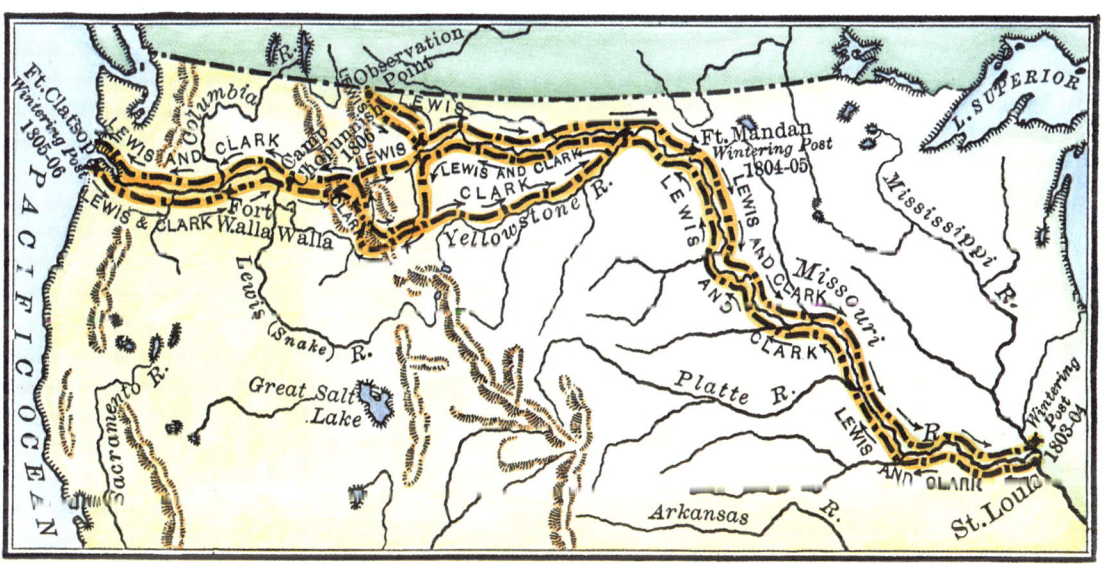

The Physician as Hero in Plague Time: Dr Benjamin Rush

Opposite: Benjamin Rush led a successful campaign against yellow fever.

Above top: Philadelphia's Independence Hall.

Above: The route taken by the Lewis and Clark expedition.

> **HUNDREDS OF PEOPLE WERE DYING EACH WEEK, AND TENS OF THOUSANDS OF PEOPLE, INCLUDING NATIONAL LEADERS, CHOSE TO FLEE THE CITY.**

more humane care. He promoted the engagement of mental patients in activities such as gardening and washing, and he was a strong advocate for the view that alcoholism is a disease. His medical students were so favourably impressed by Rush that they founded Rush Medical College in Chicago, now Rush University Medical Center, in his honour. It was Rush who persuaded former enemies Adams and Jefferson to resume their correspondence.

YELLOW FEVER

Yellow fever is caused by an RNA virus spread by the bite of infected mosquitoes. After an incubation period of several days, most patients develop mild symptoms including fever, headache, and anorexia, recovering in less than a week. But approximately 15 per cent of patients develop a second phase with recurrent fever and jaundice, from which the disease derives its name, as well as haemorrhaging from the mouth, nose, and eyes, and bloody diarrhoea. Among such patients, mortality rates may be as high as 50 per cent. Those who survive generally recover completely, with the added benefit of lifelong immunity to the disease.

When a female mosquito ingests the blood of an infected human or primate, the virus begins replicating in the epithelial cells of the creature's gastrointestinal tract. When they take up residence in the insect's salivary glands, they are then transmitted the next time the mosquito bites. Because mosquitoes tend to be active in warmer months, yellow fever outbreaks tend to occur in summer. Once the virus enters the bloodstream of a human victim, it begins reproducing in lymphatic organs, from which it can infect the cells of the liver. Cases of death are often due to the aggressiveness of the immune response, producing a vicious cycle of escalating immune response sometimes referred to as a "cytokine storm".

More than two dozen outbreaks of yellow fever marked the history of North America, involving, in addition to Philadelphia, such cities as Savannah, Georgia, New Orleans, Louisiana, and Norfolk, Virginia. Of special historical interest was the problem the disease presented to would-be builders of the Panama Canal. Initial French efforts were essentially doomed by the disease, eventuating in over 22,000 deaths, and the resulting business failure incited financial turmoil in France. The US finally succeeded in completing the canal largely through the realization of the role of mosquitoes in transmitting the disease, followed by successful efforts at eradication.

RUSH AND YELLOW FEVER

At the time of the 1793 epidemic, Philadelphia was the largest city in the United States, with a population of 50,000. It was also the US capital, and the outbreak of disease spurred efforts to move the capital to what became Washington, DC. The epidemic began in August, with the deaths of two immigrants. Rush, who had lived through another outbreak of the disease in 1762, recognized what was happening and immediately alerted officials to the return of a "highly contagious as well as mortal remitting yellow fever". Citizens were warned to avoid habits that might promote the disease, such as excessive exertion, and the city's streets were cleaned.

As the month of August wore on, the deaths of prominent citizens, including physicians active in the fight against the disease, led to increasing agitation and eventually panic. Hundreds of people were

Opposite: The mosquito that carries yellow fever and can also transmit a variety of other infectious organisms.

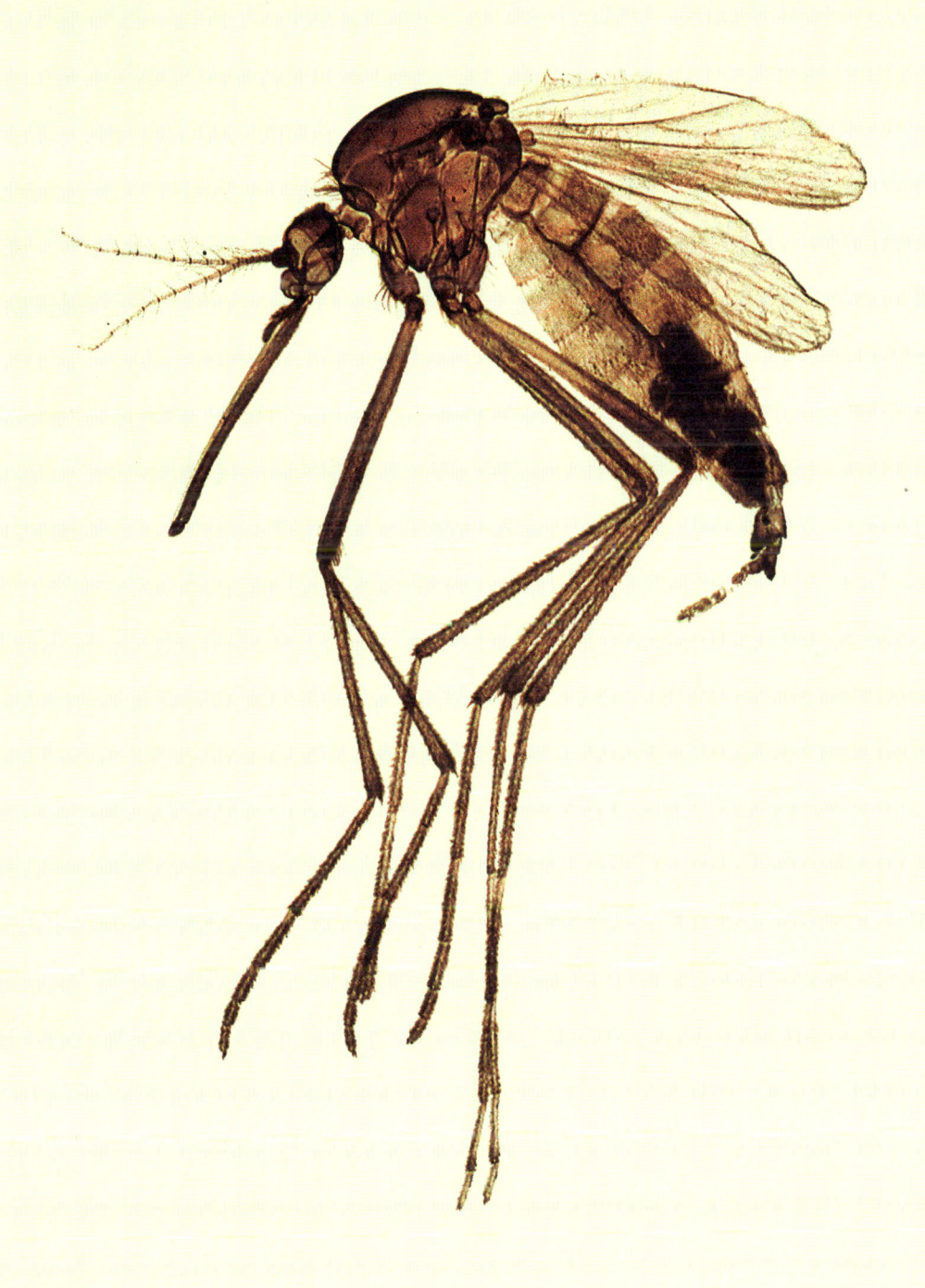

The Physician as Hero in Plague Time: Dr Benjamin Rush

dying each week, and tens of thousands of people, including national leaders, chose to flee the city. Samuel Breck, a merchant who had newly arrived in the city, described the scene:

> In private families the parents, the children, the domestics lingered and died, frequently without assistance. The wealthy soon fled; the fearless or indifferent remained from choice, the poor from necessity. The inhabitants were reduced thus to one-half their number, yet the malignant action of the disease increased, so that those who were in health one day were buried the next. The burning fever occasioned paroxysms of rage which drove the patient naked from his bed to the street, and in some instances to the river, where he was drowned. Insanity was often the last stage of its horrors.

Many physicians left the city, but Rush remained. No one, Rush included, had ever heard of a virus, and the role of mosquitoes in transmitting the disease was unsuspected by all, but some of Rush's ideas probably helped to contain the disease. He believed that the epidemic might be traced to foul vapours, and he pushed for their eradication. For example, he urged that decaying food be swept from nearby docks, that sewage be disposed of in a more sanitary matter, and he encouraged the adoption of improved hygiene. Rush resisted attempts to blame recent immigrants and instead insisted that the city be cleaned up, so that future generations would not be similarly afflicted.

A SENSE OF HUMOR

A staunch humoralist, Rush naturally treated the febrile illness with phlebotomy and purging. He also relied heavily on mercury-containing compounds of the sort he provided Lewis and Clark with for their expedition. Describing his therapeutic approach, he wrote:

> I have found bleeding to be useful, not only in cases where the pulse was full and quick but where it was slow and tense. I have bled twice in many and in one acute case four times, with the happiest effect. I consider intrepidity in the use of the lancet, at present, to be necessary, as it is in the use of mercury and jalap, in this insidious and ferocious disease.

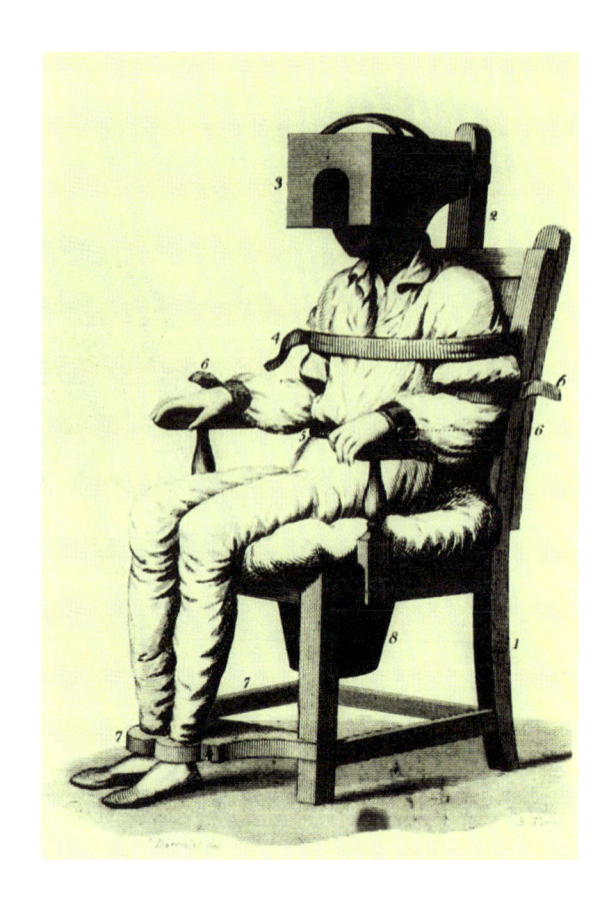

Rush recognized the limitations of the therapies available to him, but like many physicians during the 2020 COVID-19 pandemic who prescribed unproven hydroxychloroquine for patients, Rush believed that it was best to do something rather than nothing. And Rush practised what he preached. Years later, when he lay dying, he insisted on being bled himself.

A MAN OF PRINCIPLE

Perhaps Rush's greatest contribution during the yellow fever epidemic of 1793 was his moral example. At a time when many of his colleagues were fleeing the city, Rush chose to remain behind, saying, "I have resolved to stick to my principles, my practice, and my patients to the last extremity." Rush did not waver in his resolve, even though three of the apprentices he had recruited died of the disease, as did his sister. And Rush himself fell ill, for a time too sick to leave his house. Despite seeing as many as one hundred patients a day, Rush rarely failed to write to his wife about the work he was doing and sharing his prayers for his "poor patients".

As a physician with vast political experience and a deep belief in the power of institutions to improve human life, Rush naturally sought to engage existing organizations in the cause and played a prominent role in founding new ones. Each day he commuted from his house just outside the city into the new hospital that had been constructed in response to the disease, where he served as the chief physician. Believing that blacks were less susceptible to the disease, he organized groups of black women to serve as nurses, and later told his wife that the majority of his patients were cared for by "African brethren". Unfortunately, however, Rush was wrong, and blacks enjoyed no greater immunity.

Although phlebotomy and mercury-based medications may have done little to restore the health of Rush's patients, and may have even hastened the demise of some, Rush himself was celebrated as a hero, primarily because he stood by the patients and his city when many others were abandoning them. A judge wrote of Rush, "He is become the darling of the common people and his humane fortitude and exertions will render him deservedly dear." And Rush's recommendations concerning sanitation likely provided many practical benefits, undermining the conditions favourable to the propagation of other epidemic infectious diseases.

Rush's response to the yellow fever epidemic of 1793 serves as an inspiring example to health professionals confronting crises and disasters of their own. First, physicians exist to serve patients, and doing so sometimes entails personal risk. Second, by remaining calm and committed to the professional mission, physicians can serve as exemplars to others, helping them to locate their inner courage and resist their impulse to abandon their mission of service. Finally, Rush deeply understood the vital role well-led institutions could play in meeting the challenges of disaster. As a model of professional purpose and dedication, Rush summoned forth the best, both from himself and from those around him.

> **RUSH RECOGNIZED THE LIMITATIONS OF THE THERAPIES AVAILABLE TO HIM, BUT LIKE MANY PHYSICIANS DURING THE 2020 COVID-19 PANDEMIC WHO PRESCRIBED UNPROVEN HYDROXYCHLOROQUINE FOR PATIENTS, RUSH BELIEVED THAT IT WAS BEST TO DO SOMETHING RATHER THAN NOTHING.**

Opposite: Rush's tranquilizer chair, which was meant to treat mental illness.
Above: Yellow fever affected the builders of the Panama Canal.

13 EDWARD JENNER AND THE LITTLE PRICK

Edward Jenner (1749–1823) is generally credited with introducing one of the most successful means of reducing the spread of infectious disease: vaccination. In fact, as we have seen, a coordinated worldwide vaccination programme has rid humanity of what was formerly one of its most dreaded scourges, smallpox, which once took the lives of about 10 per cent of the population in European countries.

THE FIRST VACCINATIONS

Jenner was not the first person to realize that intentionally inoculating ("engrafting") non-immune individuals could produce a milder form of the disease. A lancet would be inserted into the pustule of a patient suffering from smallpox, then inserted just underneath the non-immune person's skin. This would produce a local scar, but generally provide protection against the disseminated form of the disease.

This procedure, which became known as variolation ("speckling"), was introduced into Western Europe partly through the advocacy of Lady Mary Montague (1689–1762). She travelled with her husband, the British Ambassador to Turkey, where she learned of inoculation. A legendary beauty, she had been disfigured by smallpox and her brother had died of the disease.

Determined to do everything she could to prevent the disease, she had her son inoculated while they were in Turkey, and once the family returned to England, she had her daughter undergo the procedure, as well. After the procedure was tested on prisoners, then orphaned children, two members of the royal family were inoculated, leading to its widespread acceptance.

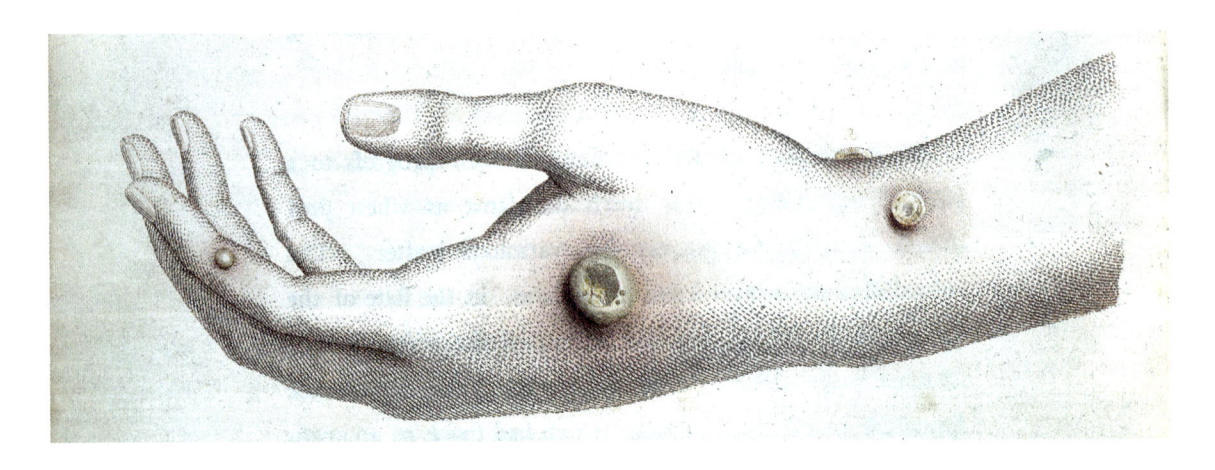

Edward Jenner and the Little Prick

SMALLPOX MORTALITY IN ENGLAND, C 1890

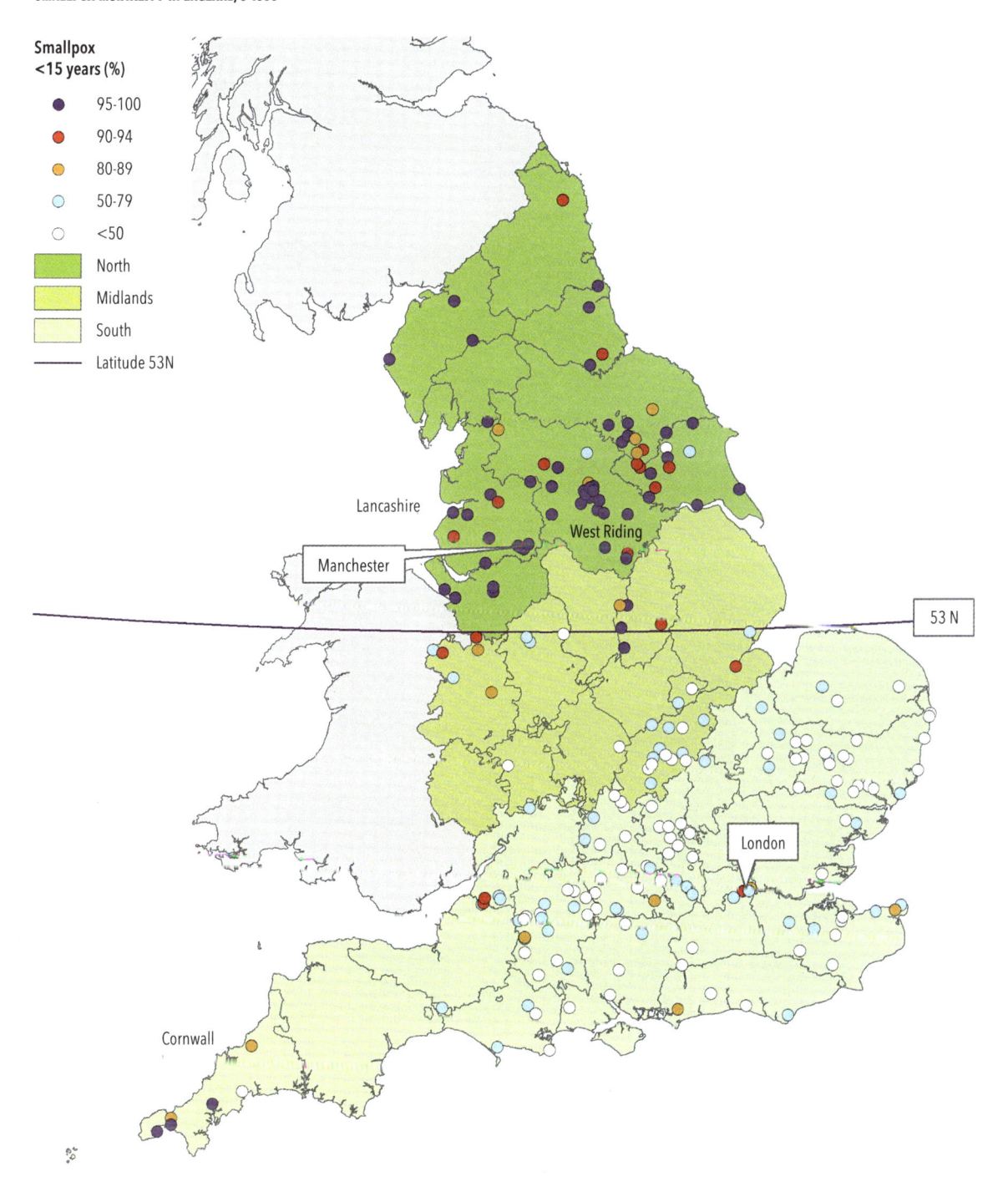

Smallpox <15 years (%)
- ● 95-100
- ● 90-94
- ● 80-89
- ● 50-79
- ○ <50

- North
- Midlands
- South
- — Latitude 53N

Lancashire
West Riding
Manchester
53 N
London
Cornwall

Opposite above: Edward Jenner infected Sarah Nelmes with the cowpox in an attempt to find a vaccine.

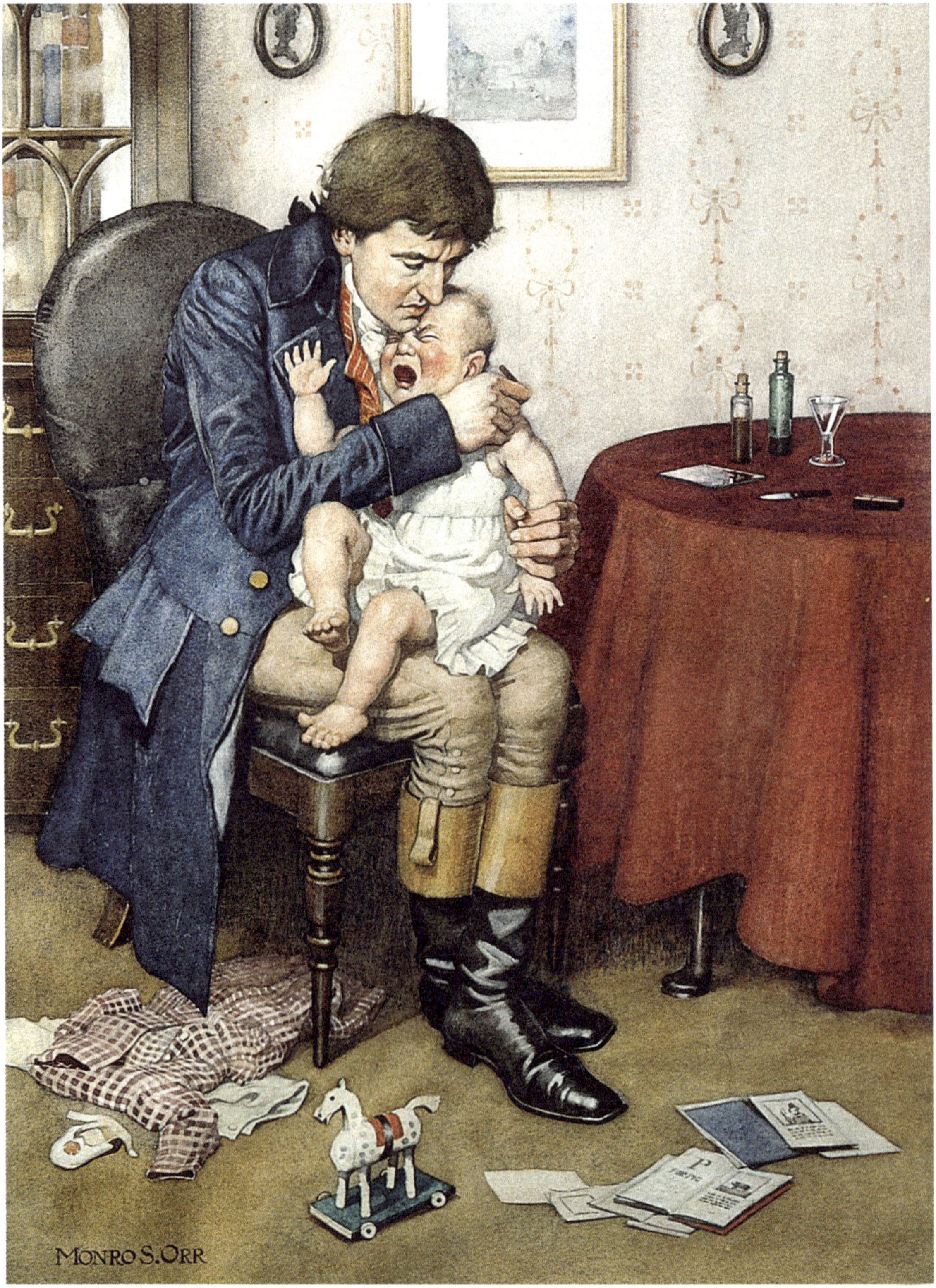
MONRO S. ORR

HE SCRAPED MATERIAL FROM THE PUSTULES ON THE HANDS OF A MILKMAID WHO HAD CONTRACTED COWPOX AND THEN INOCULATED BOTH THE BOY'S ARMS.

Variolation soon became popular, and for good reason. It is estimated that, compared to a bout of smallpox, variolation was associated with a risk of death only one-tenth as high. Because smallpox was so widespread, this appeared to many a worthwhile level of risk. However, variolation did cause death in about 2–3 per cent of patients.

PRINCIPLES OF IMMUNITY

Jenner was the eighth of his parents' nine children and underwent inoculation against smallpox as a child. Orphaned at age five, Jenner lived with his older brother and was apprenticed as a 14-year-old to a country doctor. During this time, he heard a dairymaid claim that she would never contract smallpox because she had already contracted cowpox.

Completing his apprenticeship at age 21, Jenner went to London, where he worked with one of the most famous figures in surgery, John Hunter (1728–93). Three years later, Jenner returned to the countryside, where he became a successful physician, founding a local medical society. After publishing a study on the cuckoo bird, Jenner was elected to the Royal Society in 1788.

Recalling the dairymaid's assertion that cowpox had rendered her immune to smallpox, Jenner speculated that the material from the pustules of a patient with cowpox might produce immunity from smallpox. He was not the first physician to whom this idea occurred, but he was the first to test it rigorously.

As the name suggests, cowpox is a disease of cows that can be transmitted to humans, most commonly the hands and arms of people who regularly milked cows. In fact, the disease is even more common in rodents. It is caused by the cowpox virus, which is very similar to the smallpox virus, explaining the development of cross-immunity.

In 1796, Jenner tested his hypothesis on the son of his gardener. He scraped material from the pustules on the hands of a milkmaid who had contracted cowpox and then inoculated both the boy's arms. Jenner described the procedure in these words:

The more accurately to observe the progress of the infection, I selected a healthy boy, about eight years old, for the purpose of inoculation for the cowpox. The matter was taken from a sore on the hand of a dairymaid, and it was inserted on the 14th of May, 1796, into the arm of the boy by means of two superficial incisions, barely penetrating the cutis, each about half an inch long. On the seventh day he complained of uneasiness in the axilla, and on the ninth he became a little chilly, lost his appetite, and had a slight headache. During the whole of this day he was perceptively indisposed, and spent the night with some degree of restlessness, but on the day following he was perfectly well.

After waiting nearly two months, Jenner decided to test the effectiveness of the inoculation. He did so by intentionally inoculating the boy with smallpox:

In order to ascertain whether the boy, after feeling so slight an affection of the system from the cowpox virus, was secure from the contagion of the smallpox, he was inoculated the 1st of July following with variolous matter, immediately

Opposite: Edward Jenner administering an inoculation.
Above: The process of inoculation was partly introduced to the West by Lady Mary Montague.

taken from a pustule. Several slight punctures and incisions were made on both arms, and the matter was carefully inserted, but no disease followed. Several months afterwards, he was again inoculated with variolous matter, but no sensible effect was produced on the constitution.

What made Jenner's contribution so different from others was the fact that, after inoculating patients with cowpox, he went on to intentionally challenge them with smallpox, thereby proving that the cowpox had indeed rendered them immune to the dread disease. Moreover, Jenner repeated the trial on dozens of other subjects.

REJECTIONS AND RECOGNITION

He reported his results to the Royal Society, but his findings were not accepted for publication, so in 1798 he published them himself. He continued his studies and published multiple additional papers in the years to come. He reported that he had sent cowpox material to other physicians, who had been able to replicate his results. Word of Jenner's contribution began to spread rapidly.

In 1802 Parliament awarded Jenner a prize of 10,000 pounds, and in 1807 it awarded him another of 20,000 pounds. In 1821 Jenner was appointed physician to the king, and in 1840 Parliament cemented Jenner's triumph by outlawing the practice of variolation, the practice of inoculating individuals with smallpox. Cowpox vaccination became official British policy.

Jenner's legacy lives on through the subsequent development of smallpox vaccination, which enabled the World Health Organization in 1979 to declare smallpox an eradicated disease. Of note, the word vaccination is derived from the Latin *vacca*, meaning cow. Hence the word itself represents an enduring memorial to Jenner and his work with smallpox.

Above: Edward Jenner's lancets.
Opposite: A patient suffering from smallpox.

Edward Jenner and the Little Prick

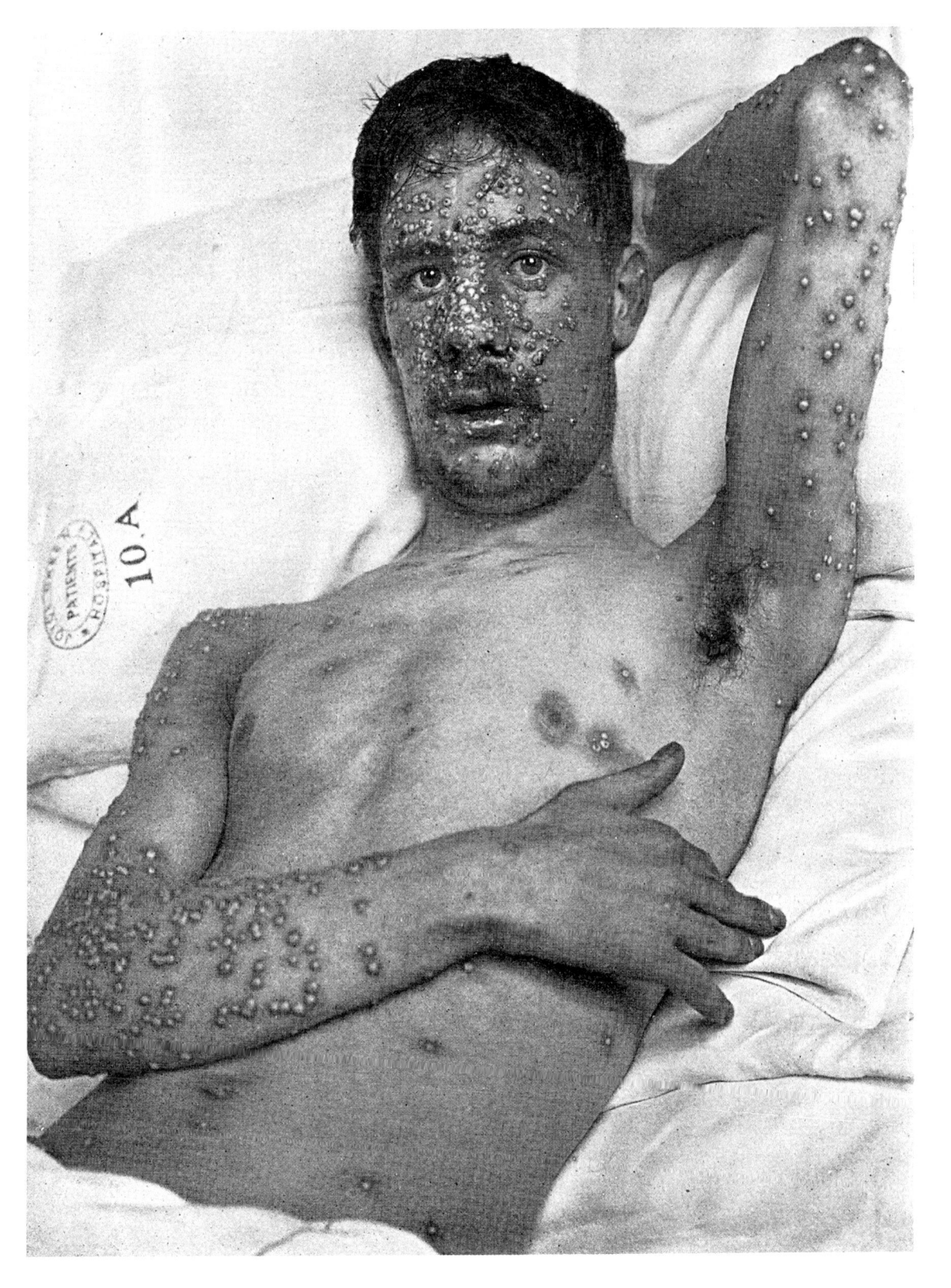

14 TUBERCULOSIS, THE PERSISTENT KILLER

Signs of tuberculosis infection, such as Pott's disease of the spine, are present in Egyptian mummies from 4,500 years ago, and written descriptions of recognizable signs and symptoms of the disease appear soon thereafter. By the second century AD, Galen had catalogued a set of symptoms including fever, night sweats, cough, and blood-tinged sputum.

By the early 1700s, physicians strongly suspected that tuberculosis was infectious, and by the nineteenth century in Western Europe, the disease was responsible for about 25 per cent of deaths. It was also during this time that the term tuberculosis was introduced and began slowly replacing the disease's former name, consumption.

Promoting the disease's rise in the eighteenth and nineteenth centuries was the industrial revolution, which led to higher population densities, crowded and poorly ventilated work and housing conditions, malnutrition, and poor sanitation standards. The pallor widely associated with the disease lent it another name, the "white plague", contrasting it with the black plague.

Today tuberculosis infects about one-quarter of the earth's human population, resulting in approximately 1.5 million deaths per year, which makes it the number-one cause of death due to a single infectious organism. More than 95 per cent of these deaths occur in so-called developing countries, such as India, China, and Indonesia.

THE SOURCE OF TB

The responsible organism is *Mycobacterium tuberculosis*, which was proved by Robert Koch to be the agent of the disease in 1882. The microbe is usually spread through the air when an infected person coughs, sneezes, or spits, and is then inhaled by an uninfected person. About 90 per cent of cases result from infection of the lungs, although other areas of the body such as the brain and bones can be involved.

Mycobacteria are unusual organisms. They divide slowly over hours rather than minutes. Their outer layer contains a much higher amount of lipid than other bacteria. And the cells that usually envelop and destroy bacteria, macrophages, are unable to digest mycobacteria, meaning that the bacilli can reproduce in such cells and eventually kill them.

Because infection is usually confined to the lungs, patients develop chest pain and a cough producing sputum. Over time, the lungs can develop scarring. Along with respiratory symptoms, patients are likely to have fever, fatigue, loss of appetite, and weight loss. The latter helps to explain the use of the term consumption to name the disease, as though the patient were being consumed.

PROGNOSIS, DIAGNOSIS, TREATMENT

Happily, about 90 per cent of patients who become infected with tuberculosis develop latent infection. This means that, while the organism is present within them, it causes no symptoms and they are not at risk of transmitting the disease to others. However, if the immune system of such patients weakens, the disease can reactivate, causing symptoms and rendering the individual infectious.

Opposite, clockwise from top left: Tuberculosis is a highly infectious disease. Until a vaccine was produced, hospitals had little choice but to keep patients in isolated wards and perform what treatments they could.

71

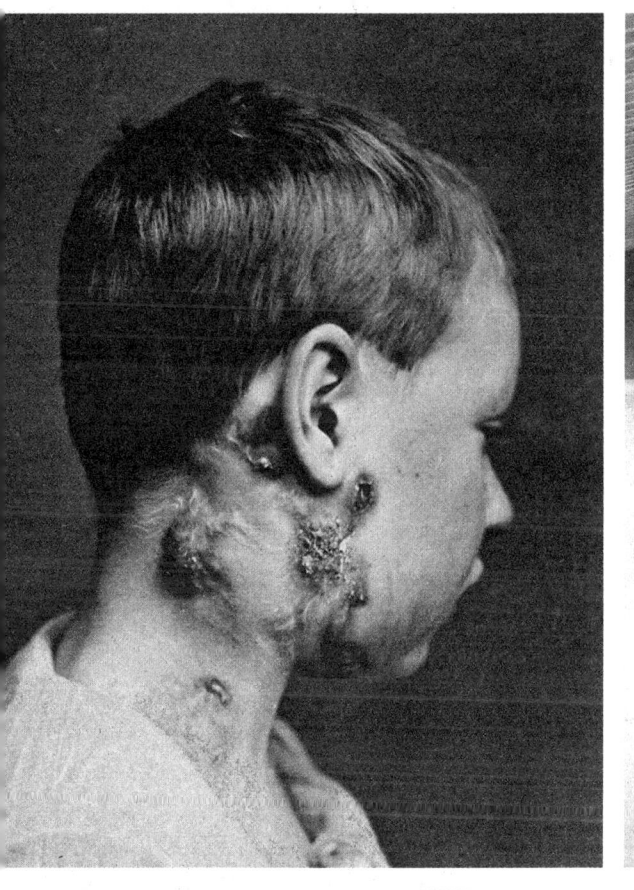

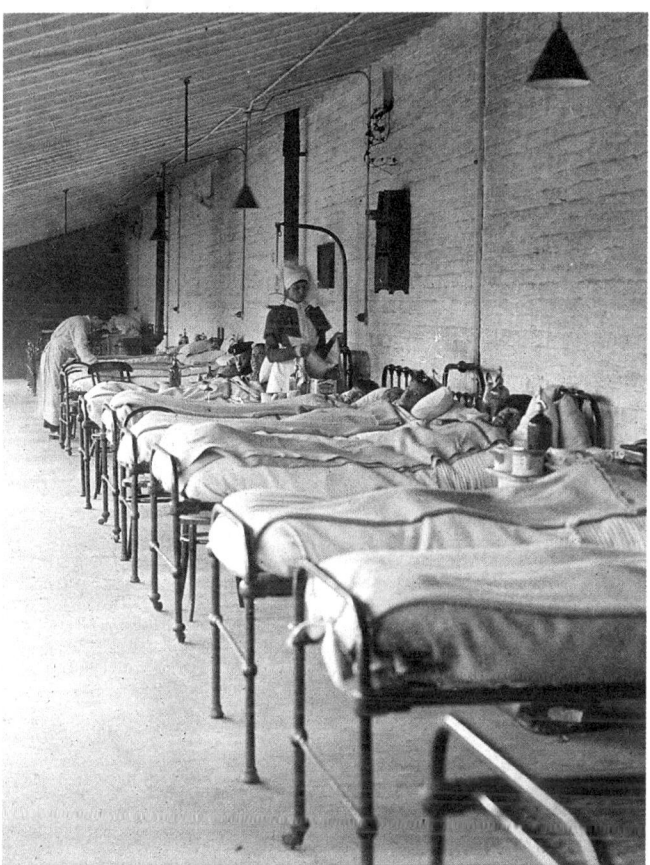

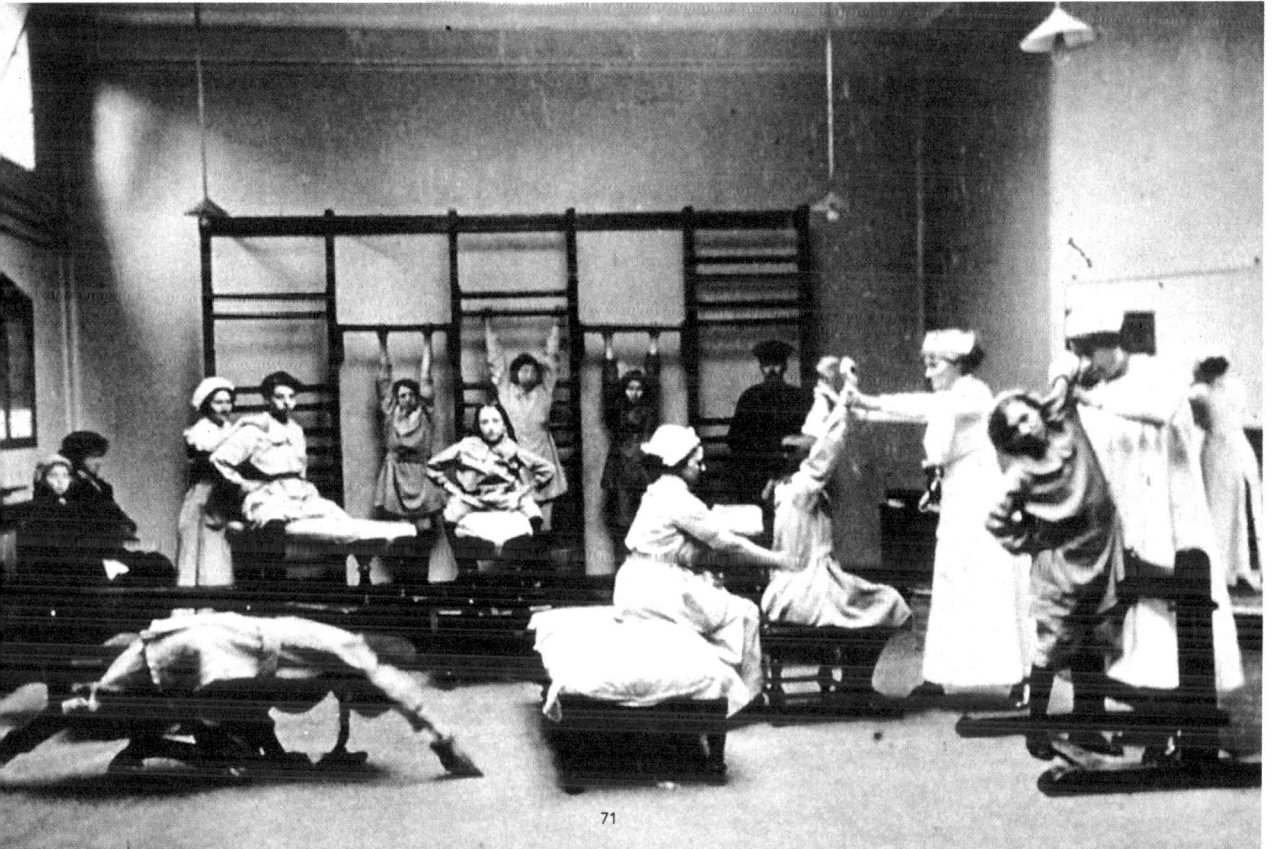

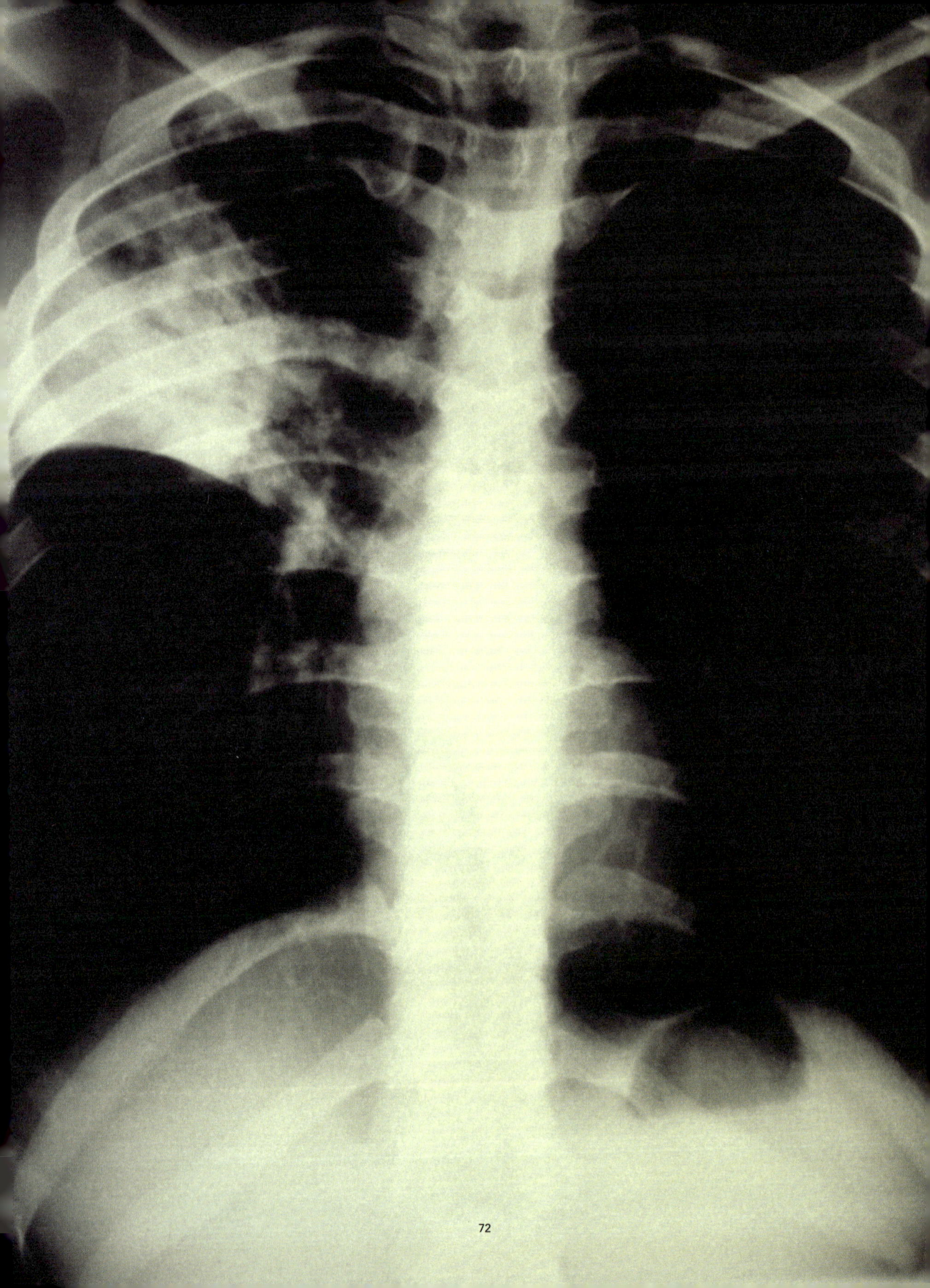

Active tuberculosis can be diagnosed when the organism grows out of cultures of sputum or other fluids or tissues. One major drawback of this diagnostic approach is the fact that, because the bacillus divides so slowly, it can take weeks to obtain a positive result. For this reason, treatment is often commenced before a definitive diagnosis is obtained.

In many parts of the world, the BCG vaccine is used to contain tuberculosis. While far from 100 per cent effective, it can reduce the probability of becoming infected by about 20 per cent and prevent latent disease from becoming active in more than half of patients. However, it also makes tuberculin skin tests positive, which means that it is used mainly in the developing world.

Only about 10 per cent of people with latent tuberculosis ever develop active disease, and such reactivation typically occurs years to decades after initial infection. Yet if the immune system is damaged, for example in HIV/AIDS, the risk of reactivation increases markedly. In patients infected with HIV, the activation rate goes up to about 10 per cent per year, with relatively high rates of mortality.

Treatment of TB aims to eradicate the organism from the infected patient, to prevent death from infection, and to decrease transmission of the organism to other individuals. While standard drugs kill most of the organisms in the first two months of treatment, its slow growth rate and presence in macrophages can protect some organisms for longer periods of time. As a result, therapy often lasts six months.

If all the mycobacteria are not killed, the infection can recur, causing the patient to become both ill and infectious again. To eradicate the organism requires multiple drugs that attack it in different ways, decreasing the probability that resistance to treatment will develop. A typical regimen consists of four drugs: isoniazid, rifampin, pyrazinamide, and ethambutol.

Effective treatment is difficult for many reasons. First, patients need to take multiple drugs for an extended period of time, long after any symptoms may have abated. Second, the disease is most common in poor parts of the world, where follow-up can be difficult. For this reason, patients are often directly observed taking their medications, to verify that they are doing so correctly.

> **THE MICROBE IS USUALLY SPREAD THROUGH THE AIR WHEN AN INFECTED PERSON COUGHS, SNEEZES, OR SPITS, AND IS THEN INHALED BY AN UNINFECTED PERSON.**

Opposite: A chest radiograph shows abnormal opacity in the upper lobe of the right lung in a patient with tuberculosis.
Above: A public information poster warning of the dangers of contagion.

WORLDWIDE DISTRIBUTION OF TUBERCULOSIS CASES

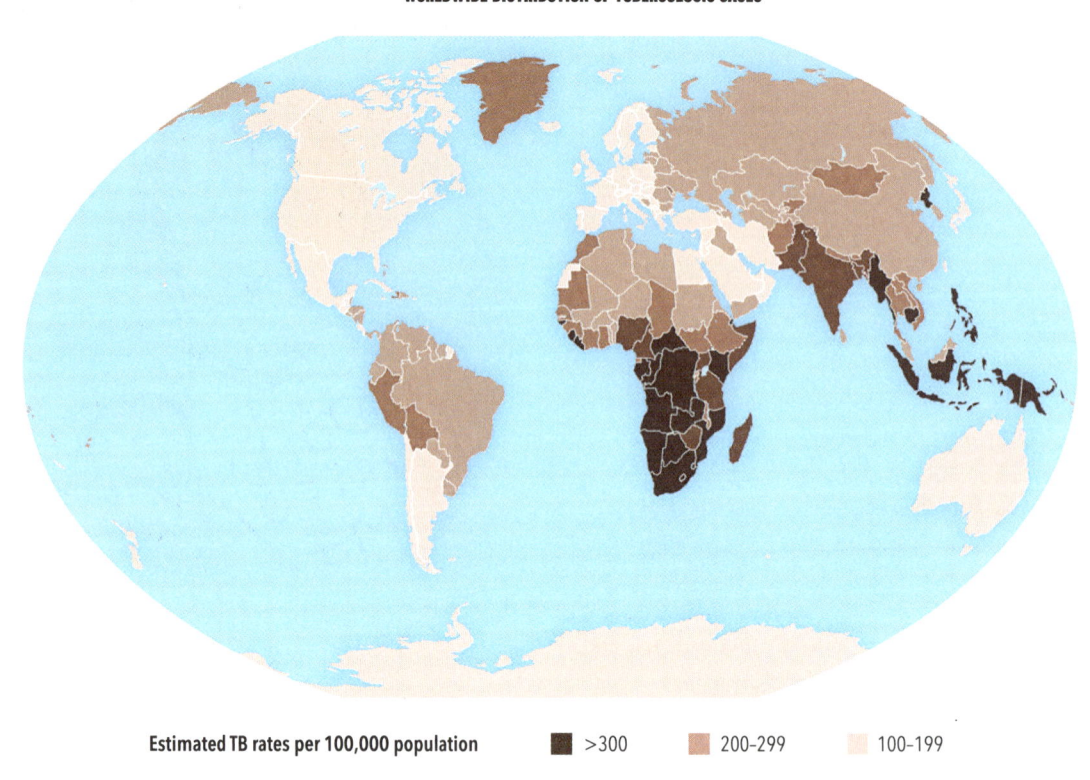

Estimated TB rates per 100,000 population ■ >300   ■ 200–299   ■ 100–199

Unfortunately, strains of the bacillus resistant to multiple drugs have developed in numerous parts of the world, in part due to incomplete or inadequate treatment. Treating such infections requires more expensive medication, and some strains of the mycobacterium have been isolated that are resistant to all drugs currently in use.

While drug treatment is vital, infection and death rates due to tuberculosis were falling rapidly in many countries before the introduction of effective anti-tuberculous drugs in the twentieth century. Key factors included improved nutrition, living conditions, and sanitation – bolstering host resistance and reducing transmission of the organism.

Right: TB treatments today are far more effective.
Opposite, from top: The bacterium that causes tuberculosis is under far better control today thanks to vaccines, but cases have recently started to rise.

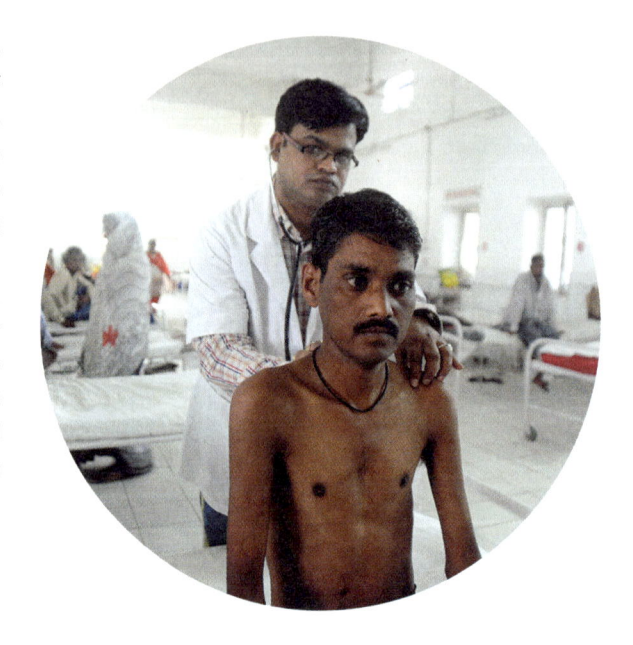

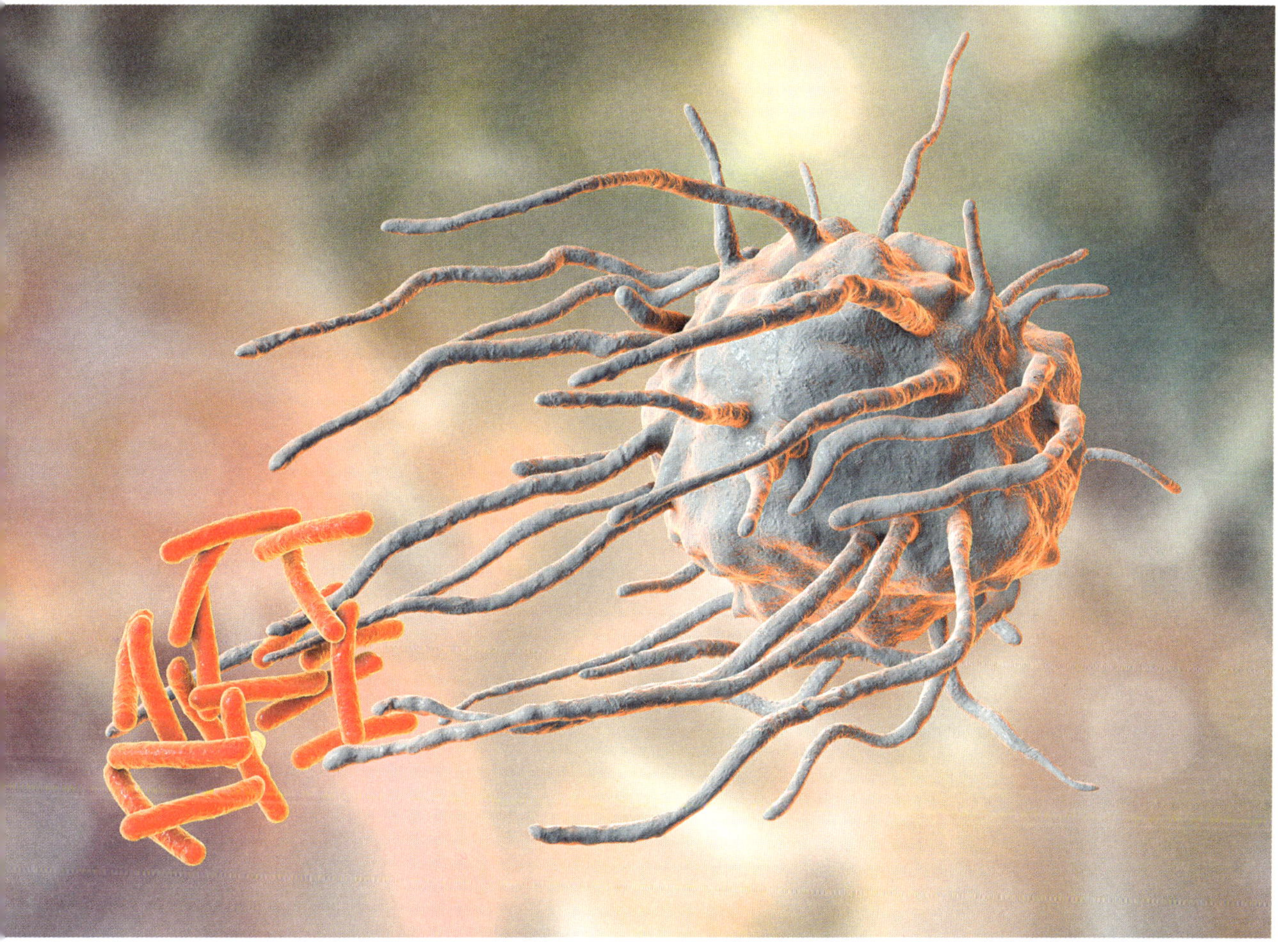

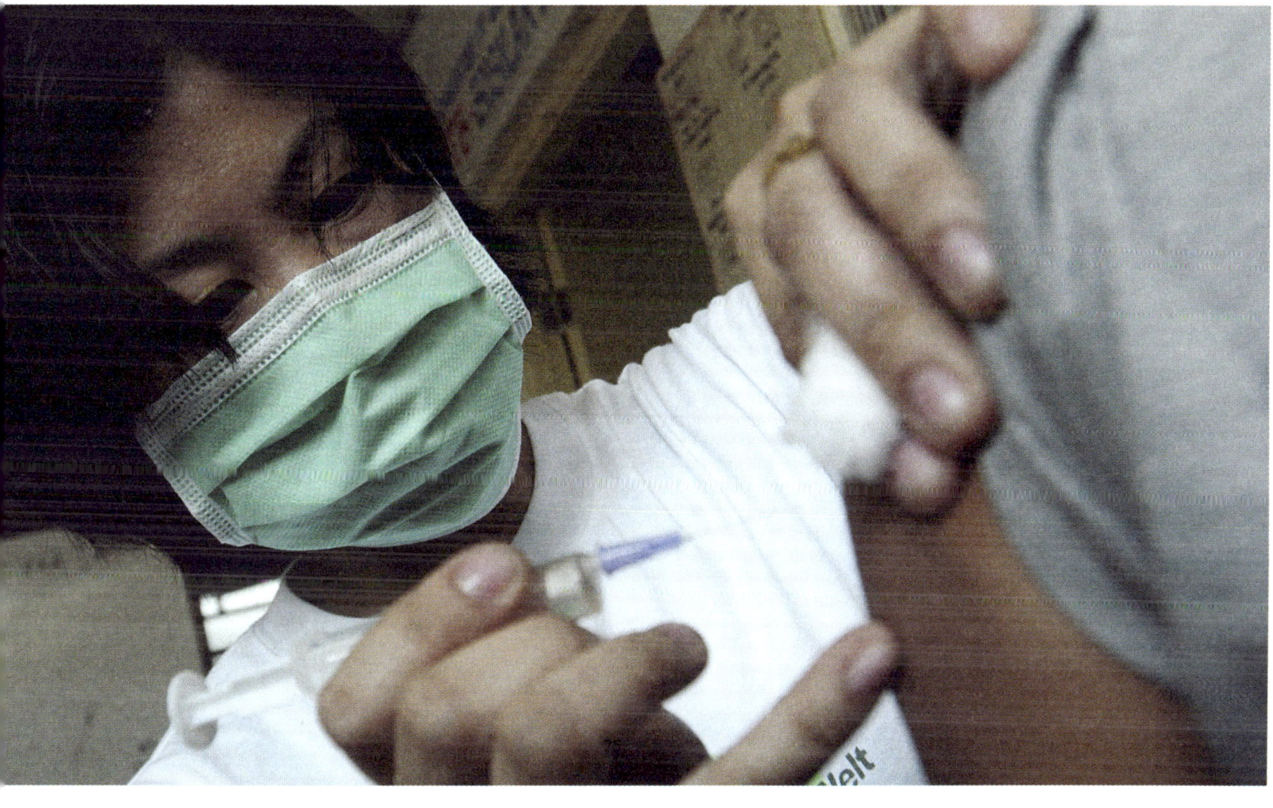

15

TUBERCULOSIS: A POETIC CASE

It is possible to learn a good deal about a disease by studying it in the abstract, in terms of its stereotypical symptoms and signs, clinical course, causes, diagnosis, and treatment. But when students enter the study of medicine, they learn not only from textbooks and journal articles, but also from clinical practice – in other words, from the care of particular patients.

The highly respected American physician Sir William Osler famously said that "the student begins with the patient, continues with the patient, and ends his study with the patient", using all the other resources of medical education "as means to an end". The same applies to the study of the history of infectious disease – we can learn a good deal by studying diseases, but we also need to become acquainted with the stories of patients.

To this end, we turn now to the story of a particular patient, a man who stands alongside Lord Byron (1788–1824) and Percy Shelley (1792–1822) as one of the three great romantic poets and is often considered one of the most beloved of all English poets, in some estimations second only to Shakespeare.

This poet is John Keats (1795–1821), and his genius as a poet is made even more remarkable by the fact that, in addition to completing his training in medicine, which enabled him to diagnosis his own eventually fatal illness, he produced essentially all of his poetry in the short space of four years and died of tuberculosis at the age of only 25 years.

FAMILY AND COURTSHIP

Keats was born in London, the son of an innkeeper who died when he was a boy. His mother died of tuberculosis when he was only 14 years old. Keats undertook the study of medicine at Guy's Hospital in 1815, completing his studies a year later, but upon graduation he intended to be a poet, not a surgeon.

Always a lover of literature, Keats published his first volume of verse in late 1816. He made friends with a number of prominent literary figures, some of whom promoted his work tirelessly. Keats and his brother George tended their brother Tom, who was suffering from tuberculosis. In 1918, Keats commenced a walking tour and George emigrated to the United States. Tom died late that year.

Above: William Osler (1849–1919).
Right: Percy Bysshe Shelley (1792–1822).
Opposite: John Keats (1795–1821).

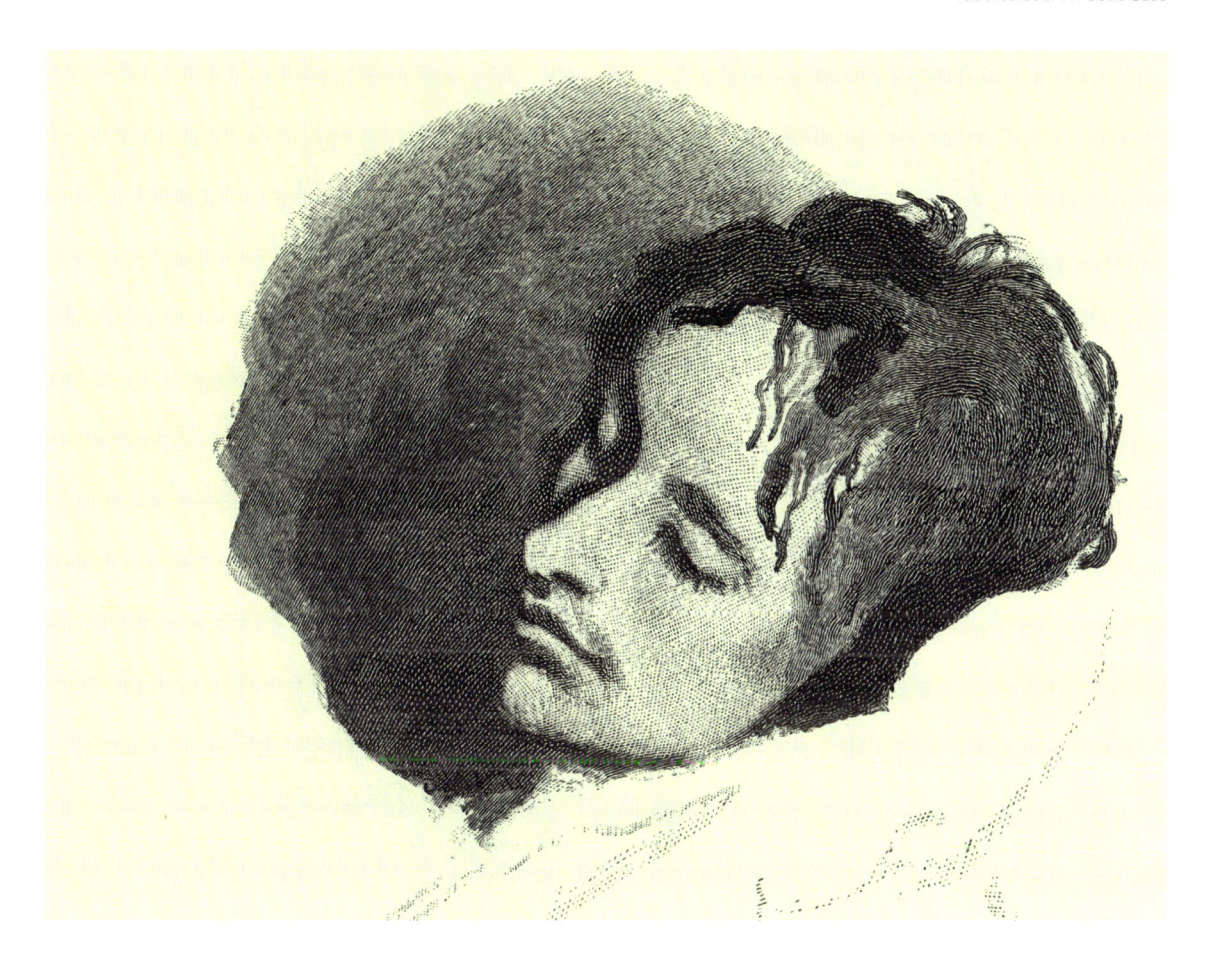

After his brother's death, Keats moved into a house owned by a friend and began a period of remarkable creativity, composing such works as "Ode to a Nightingale" and "Ode on a Grecian Urn", the latter ending with the memorable lines, "Beauty is truth, truth beauty – that is all ye know on earth and ye need to know."

Late in 1818 Keats also met Fanny Brawne, who lived nearby with her mother, and the two seem to have fallen in love. Keats wrote her many letters, and they may have even planned to be married, although his uncertain professional and financial prospects at this point seem to have prevented any sort of formal engagement.

THE CONSUMPTIVE ARTIST

Perhaps in part because Keats had nursed his mother and brother during their battles with tuberculosis, Keats himself began showing signs of tuberculosis. At the time, this name had not yet been coined, and the disease was generally known as phthisis, the Greek word for consumption. A wasting disease, it was associated with fever, cough, and sometimes haemoptysis, the coughing up of blood.

A friend described a night in 1820, when Keats had just returned to the house:

He arrived at 11:00 in the night, in a state that looked like a fierce intoxication. As such a state in him I knew was impossible, it therefore was the more fearful. I asked hurriedly, "What is the matter? You are fevered."

"Yes, yes," he answered. "Fevered, of course, a little."

I followed with the best immediate remedy in my power. I entered his chamber as he leapt into bed. On entering the cold sheets, before his head was on the pillow, he slightly coughed, and I heard him say, "That is blood from my mouth."

I went toward him; he was examining a single drop on the sheet.

"Bring me the candle, Brown; let me see this blood," and after regarding it steadfastly, he looked up in my face, with a calmness of countenance that I can never forget, and said, "I know the colour of that blood; it is arterial blood. I cannot be deceived in that colour; that drop of blood is my death warrant; I must die."

Keats's letters to Fanny provide numerous accounts of his condition. Early in 1820, he writes:

Indeed I will not deceive you with respect to my health. This is the fact as far as I know. I have been confined three weeks and am not yet well – this proves there is something wrong about me which my constitution will either conquer or give way to.

By late summer of the same year, he had become considerably bleaker in the assessment of his prospects:

I am sickened at the brute world which you are smiling with. I see nothing but thorns for the future. I see no prospect of any rest. I wish you could infuse a little confidence in human nature into my heart. I cannot muster any – the world is too brutal for me – I am glad there is such a thing as the grave – I am sure I shall never have any rest until I get there.

Advised by his physicians that his only hope lay in a change of climate, in September of 1820 Keats and a friend, Joseph Severn, sailed to Italy, settling in Rome. There a local physician, applying ancient humoralist principles, prescribed a regimen of bloodletting and sharply reduced diet, intended to relieve his fever. Probably at least in part due to this treatment, Keats's condition steadily worsened.

Severn, who attended Keats, wrote of his desperation at his friend's decline,

Little did I think what a task of affliction and danger I had undertaken, for I thought only of the beautiful mind of Keats, [and] my attachment to him. He remains quiet and submissive under his heavy fate. For three weeks I have never left him. I have nothing to break this dreadful solitude but letters. Day after day, night after night, here I am by our poor dying friend. My spirits, my intellect and my health are breaking down. I can get no one to change with me – no one to relieve me. All run away, and even if they did not Keats would not do without me.

Above: Fanny Brawne (1800–65).
Opposite: Keats's death mask. Keats has come to be seen as the stereotypical consumptive poet.

DEATH AND LEGACY

Keats died on 23 February 1821. He is buried in the Protestant Cemetery in Rome, very near the Spanish Steps, where, according to his instructions, the gravestone bears these words: "Here lies one whose name was writ in water." Seven weeks after Keats's death, Shelley wrote his poem *Adonais*, evoking the tragedy of his friend's death:

The loveliest and the last,
The bloom, whose petals nipped before they blew
Died on the promise of the fruit.

Keats died plagued by the fear that his life had amounted to nothing. In a letter to Fanny in September of 1819, he wrote: "'If I should die,' I said to myself, 'I have left no immortal work behind me – nothing to make my friends proud of my memory – but I have loved the principle of beauty in all things, and if I had had time I would have made myself remembered.'" On this point, at least, Keats was wrong, and although the disease we know as tuberculosis robbed him of life at a young age, the beauty of his poetry and letters continues to shine.

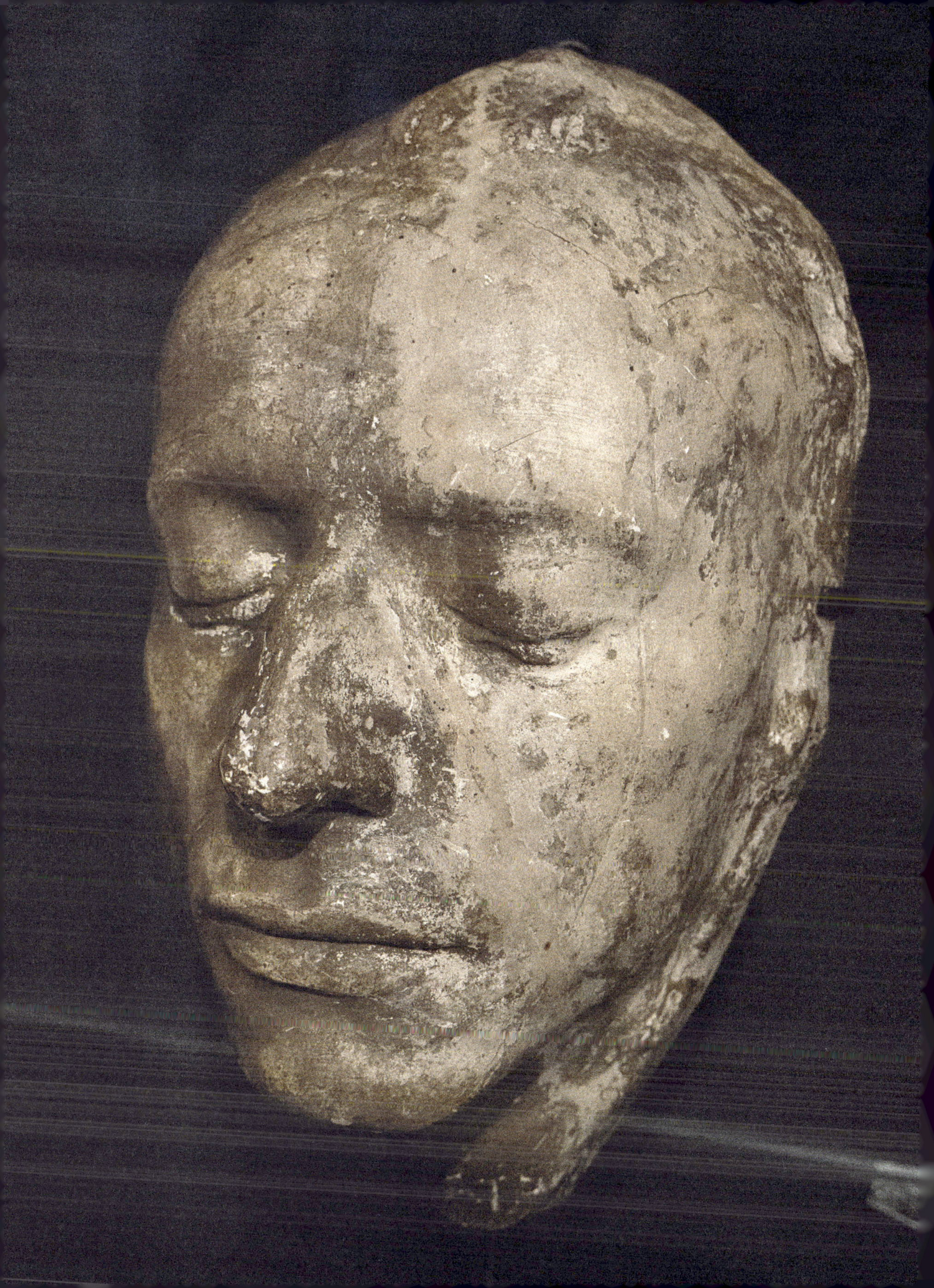

16 JOHN SNOW, FOUNDER OF EPIDEMIOLOGY

John Snow (1813–58) was born the eldest of nine children of a labourer and grew up in poverty. From a young age, he displayed a gift for mathematics and he started a medical apprenticeship when he was just 14. He was admitted to the Royal College of Surgeons in 1838 and the Royal College of Physicians in 1850, and in the same year became a founding member of the Epidemiological Society of London.

OBSTETRICS

As an obstetrician, Snow pioneered the study of both ether and chloroform as inhalation anaesthetics, designing devices to permit their controlled and safe administration. His work in the field was so highly esteemed that Queen Victoria elected to give birth to her final two children under anaesthesia, which Snow himself administered – a widely influential implicit endorsement of its use in childbirth and surgical procedures, which helped to establish anaesthesiology in England.

EPIDEMIOLOGY

Snow also helped to found the discipline of epidemiology, thanks to his painstaking investigation and analysis of a cholera outbreak in the London area of Soho. His ideas introduced public health reforms in water and waste systems that dramatically reduced the incidence of infectious diseases.

The term epidemiology, the study of epidemics, is derived from the Greek roots *epi-*, meaning "upon", and *-demos*, for "people". Today, epidemiology is the study of the distribution of diseases among people, places, and times, including the identification of risk factors and causes of disease.

SNOW AND THE BATTLE AGAINST CHOLERA

During his time as a medical apprentice, Snow encountered his first epidemic of cholera, a bacterial disease of the intestines characterized by profuse watery diarrhoea that can rapidly progress to death due to dehydration. It is contracted by a faecal–oral route through water and food.

Tracing and Mitigation

Snow is best known today for his investigation into an 1854 outbreak of cholera in Soho. At the time, in-home running water and toilets were rare, and most people got water for drinking, cooking, bathing, and washing from a communal pump. Moreover, sewage was often dumped into the streets, into open pits known as cesspools, or into the River Thames. Many authorities believed that cholera was caused by breathing foul vapours.

Snow was convinced that cholera could be traced to contaminated water, and the 1854 Soho outbreak, which killed over six hundred people, provided

Above: John Snow.
Opposite: Public health announcement concerning contamination of drinking water.

John Snow, Founder of Epidemiology

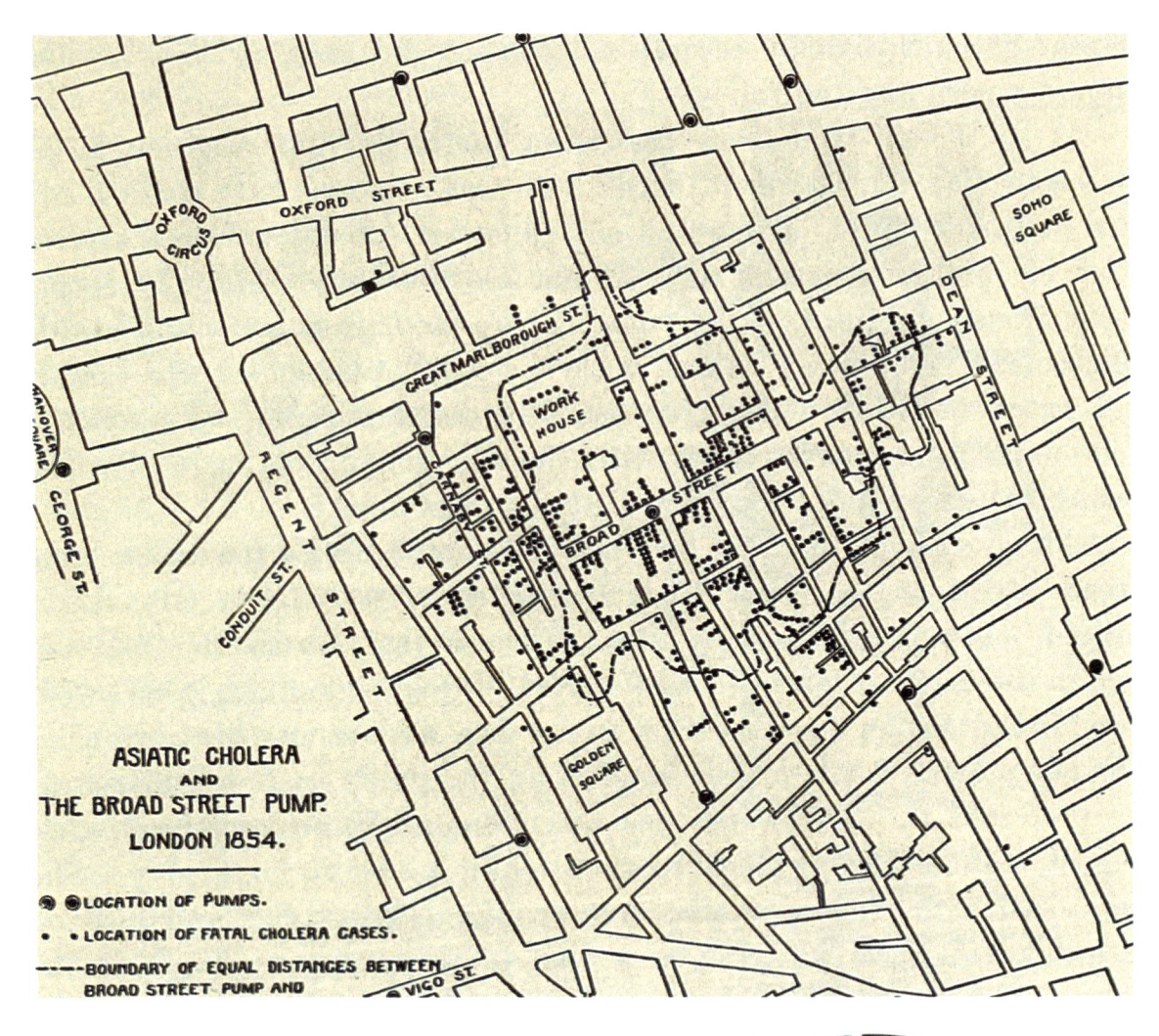

ASIATIC CHOLERA AND THE BROAD STREET PUMP. LONDON 1854.

- ● LOCATION OF PUMPS.
- • LOCATION OF FATAL CHOLERA CASES.
- --- BOUNDARY OF EQUAL DISTANCES BETWEEN BROAD STREET PUMP AND

an opportunity to put his hypothesis to the test. Describing his response to the outbreak, Snow wrote:

> Within 250 yards of the spot where Cambridge Street joins Broad Street there were upwards of 500 fatal attacks of cholera in 10 days. As soon as I became acquainted with the situation and extent of this eruption, I suspected some contamination of the water of the much-frequented street pump in Broad Street.

Snow devoted himself almost entirely to determining the residence of each of the cholera victims. His technique was simple but revolutionary. On a street map of the area, he placed a dot for the residence of each cholera case. It immediately became clear that the cases were indeed clustered around the Broad

> **SNOW'S IDEAS INTRODUCED PUBLIC HEALTH REFORMS IN WATER AND WASTE SYSTEMS THAT DRAMATICALLY REDUCED THE INCIDENCE OF INFECTIOUS DISEASES.**

Street pump. He even showed that other clusters of cases seemingly removed from the pump's vicinity could also be traced to it. He wrote:

> There were only 10 deaths in houses situated decidedly nearer to another street pump. In five of these cases the families of the deceased persons informed me that they always sent to the pump in Broad Street, as they preferred the water to that of the pumps which were nearer. In three other cases, the deceased were children who went to school near the pump in Broad Street.

Snow took his investigation even further. He investigated the water sources of people who did not develop cholera and showed that they were getting their water from different pumps or from their own wells. In short, there appeared to be very clear positive correlation between the incidence of cholera and getting water from the Broad Street pump, while those who got their water from other sources were much less likely to develop the disease.

Snow took his findings to local parish officials and explained them. The very next day, they arranged to have the handle of the Broad Street pump removed, rendering it inoperable. Over the succeeding days and weeks, the number of new cases of cholera began dropping, and soon residents who had left the area for fear of contracting the disease began to return. Snow's intervention seemed a great success, although he could not identify the source of the contamination.

Years later, investigators discovered that the Broad Street pump had been installed just feet away from an old cesspit that had begun to leak faecal matter into the water. The pump's proximity to the cesspit had been overlooked because the street had been widened, covering up any trace of it. Snow did not have the benefit of this knowledge, which would have significantly strengthened his case.

The aftermath

Although Snow's efforts appeared to have been successful, officials refused to believe his explanation of the outbreak. Perhaps they found the notion of faecal–oral transmission simply too distasteful to entertain. Or perhaps they were reluctant to consider the costs that would be required to improve drinking water supplies and sewage disposal. It would take the work of other investigators, such as Robert Koch, before the powers that be would finally be convinced to make the necessary investments.

CHOLERA TODAY

Despite Snow's contributions, cholera remains a major public health problem in some parts of the world today. It is estimated that several million people globally contract the disease each year, resulting in as many as one hundred thousand deaths. Unsurprisingly, infection and death rates are highest in poor areas. The disease is treated by oral rehydration therapy, the replacement of body fluids with a solution containing sugar and salts.

PERSONAL LIFE, DEATH, AND LEGACY

Snow's personal life was unusual. He adopted vegetarianism as a young man and for most of his life eschewed the consumption of alcoholic drinks. Practising what he preached, he also avoided drinking water that had not been boiled. And he never married. He suffered a stroke and died at the age of only 45 years. Today, the John Snow Society works to promote his life and works, and they sponsor an annual Pumphandle Lecture series in epidemiology in his honour.

Opposite, from top: Snow's map of the residences of cholera victims; a replica of the Broad Street pump.

17 IGNAZ SEMMELWEIS, APOSTLE OF HANDWASHING

Ignaz Semmelweis (1818–65) was a Hungarian physician who demonstrated that maternal deaths associated with childbirth could be reduced through handwashing. Despite his contribution, which led some to dub him "saviour of mothers", Semmelweis was roundly criticized and came to an early and unfortunate end.

Born in what is today part of Budapest to a prosperous merchant, Semmelweis received his medical degree in 1844 and embarked on a career in obstetrics. While working in the Vienna General Hospital, he observed that the maternal mortality rate in one of the two obstetrical clinics was only 40 per cent that of the other, and the rate was still lower for women who gave birth before reaching the hospital.

The deaths were due to a disease known then as "child bed fever". Today, we know this to be caused by the growth of bacteria in the female reproductive tract following a birth or miscarriage. Patients develop high fever and may have an abnormal vaginal discharge. In Semmelweis's day, however, "germs" were not yet accepted as the cause of disease.

Searching for an explanation, Semmelweis observed that the clinic with the lower rate of death was staffed by midwives, while medical students staffed the deadlier unit. He considered multiple possible explanations, such as overcrowding and climate, but concluded there must be some other cause. Finally, he discovered that a colleague had died of a disease very much like the one killing obstetrical patients, following a puncture by a medical student's scalpel during a procedure.

How could a man with a puncture wound die of the same disease as the obstetrical patients? Semmelweis hypothesized that "cadaverous material" carried by students who had just performed an autopsy might be carried into the delivery rooms, and that similar particles had caused the death of his colleague. Medical students participated in autopsies, but midwives did not.

Semmelweis reasoned that hand washing might remove the cadaverous material. So he began requiring medical students to wash their hands in a chlorinated lime solution before any deliveries. He chose chlorine because it proved most effective in

Above: Ignaz Semmelweis.
Right: Streptococci, the type of bacteria responsible for childbed fever.

eliminating the putrid odour of the autopsies. To his delight, his policy reduced the maternal mortality rate to 10 per cent of its former level.

It struck Semmelweis's colleagues as absurd that mere handwashing could produce such a dramatic reduction in mortality rates. Instead of hailing him as a hero, they ridiculed him and eventually dismissed him from his position. Semmelweis began assailing them in open letters, describing naysayers as murderers. His colleagues concluded that he was losing his mind and had him committed to a lunatic asylum.

Possibly as the result of injuries he received from a beating by his guards, Semmelweis developed an infection in his hand and died just two weeks later from the same high fever and prostration – an infection of the blood now referred to as sepsis – that had once taken the lives of so many of his patients. Only decades later would the germ theory of disease become widely accepted.

Right: Semmelweis's treatise on the prevention of childbed fever.

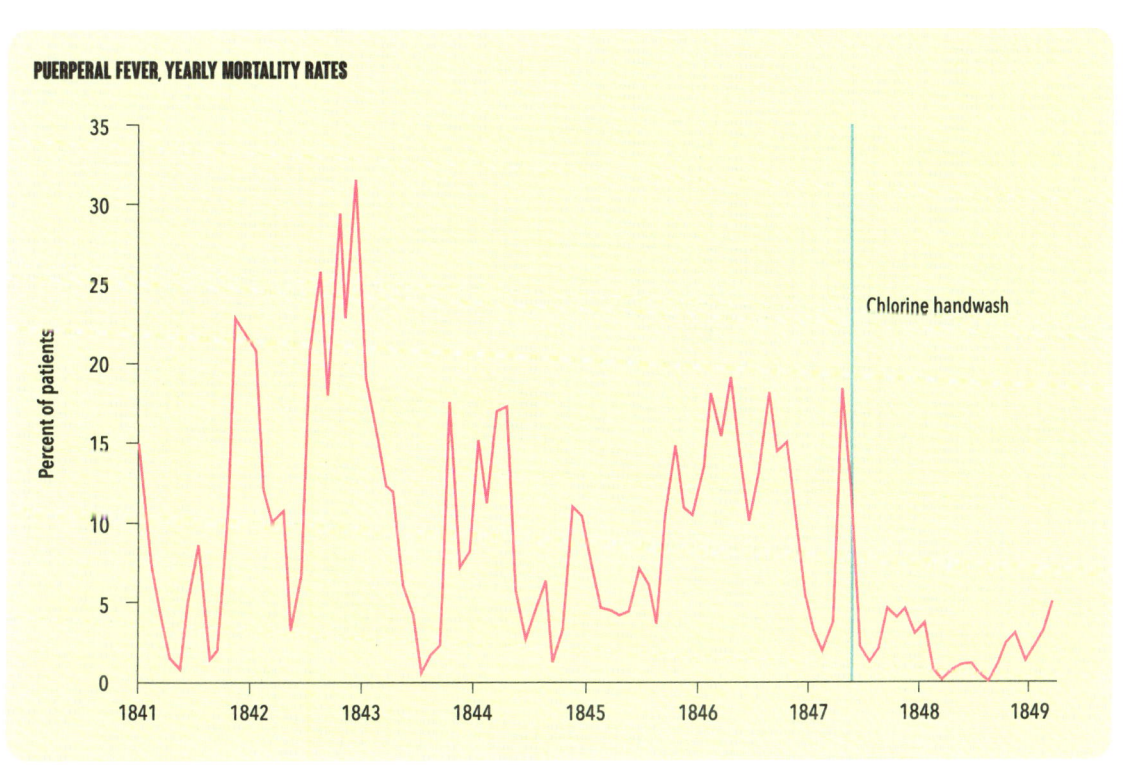

PUERPERAL FEVER, YEARLY MORTALITY RATES

18 JOSEPH LISTER, MICROBE KILLER

Joseph Lister (1827–1912) was a British surgeon who introduced antisepsis, the use of chemicals now known as antiseptics, to reduce the rate of surgical infections. Although Lister developed relatively few new operative techniques, his reductions in the rate of surgical infection and death earned him the sobriquet of "the father of modern surgery".

Lister was born in Essex, the son of a Quaker wine merchant who helped to develop the compound microscope. He studied medicine at University College London, then trained in surgery. His wife Agnes, the daughter of a prominent Scottish surgeon, also became his lifelong partner in laboratory research. While working in Glasgow, Lister read a paper by Louis Pasteur, which argued that food spoilage was due to the growth of microorganisms.

FIGHTING THE MICROBES

Pasteur suggested that chemicals could be used to kill such microbes. Lister was aware that a scientist had discovered that phenol (then known as carbolic acid) could similarly be used to prevent sewage from generating foul odours, so Lister tried spraying it on surgical instruments and wounds. When he applied it to the leg of a boy with an open leg fracture, he found that infection never developed. He published his results in *The Lancet*, writing:

> Bearing in mind that it is from the vitality of the atmospheric particles that all the mischief arises, it appears that all that is requisite is to dress the wound with some material capable of killing these septic germs, provided that any substance can be found reliable for this purpose, yet not too potent as a caustic. In the course of the year 1864 I was much struck with an account of the remarkable effects produced by carbolic acid upon the sewage of the town of Carlisle, the admixture of a very small proportion not only preventing all odour from the lands irrigated with the refuse material, but, as it was stated, destroying the entozoa which usually infest cattle fed upon such pastures.

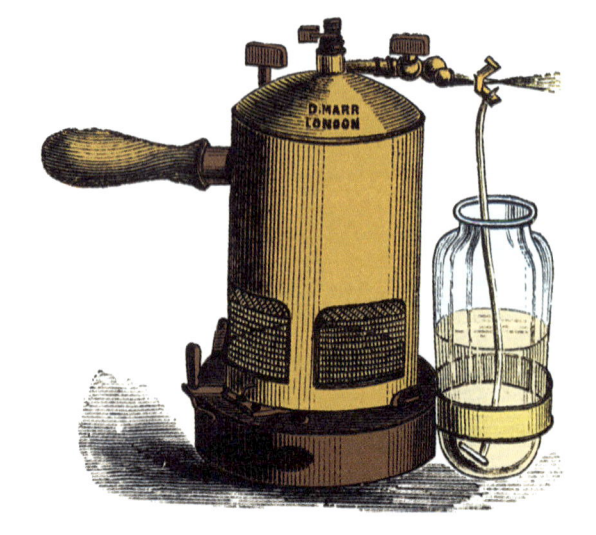

Right: An early antiseptic sprayer.
Opposite: Joseph Lister; a microscope used by Lister.

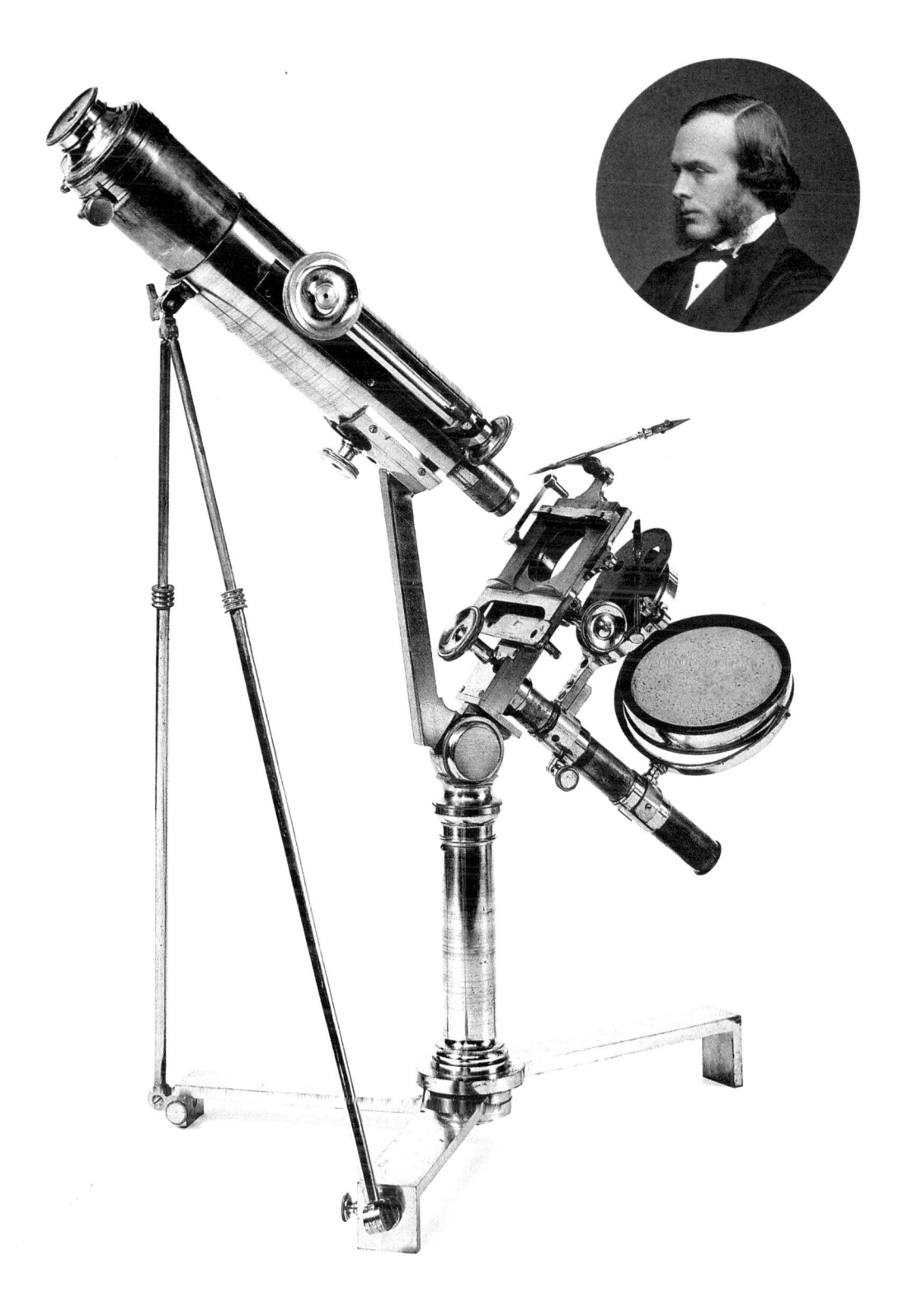

Joseph Lister, Microbe Killer

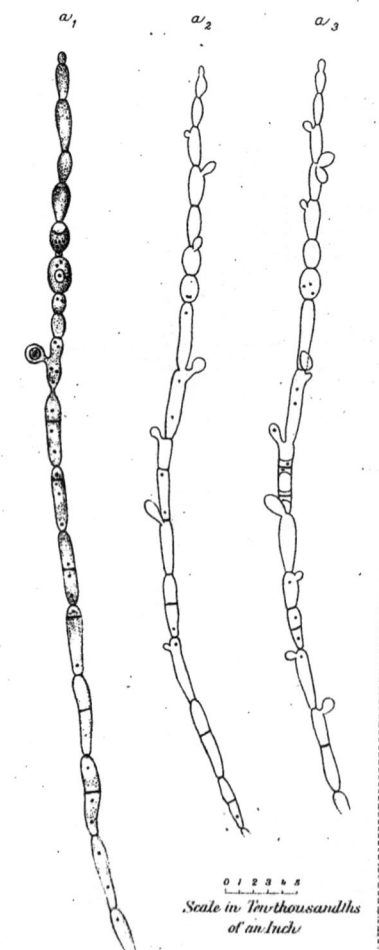

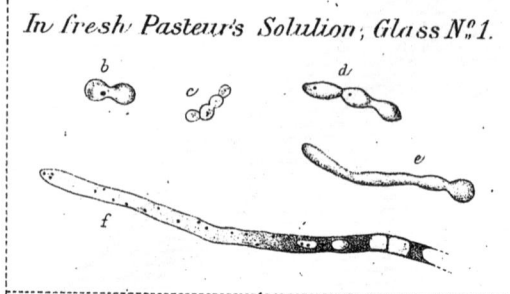

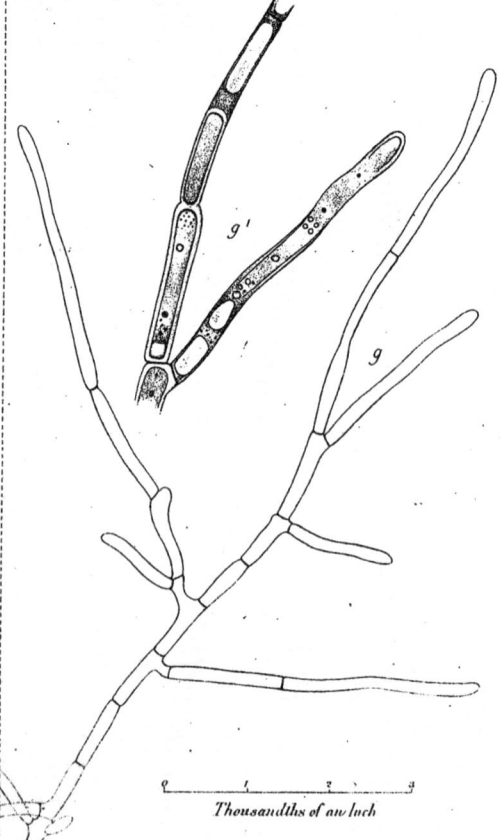

Oidium Toruloides

Fructifying Filament From a glass of stale Pasteur's Solution examined in Water.

15th Aug. 7.25 p.m. — 16th Aug. 11.45 a.m. 1.45 p.m.

a_1 a_2 a_3

Scale in Ten thousandths of an Inch

In fresh Pasteur's Solution, Glass N.° 1.

b, c, d, e, f

In fresh Pasteur's Solution, Glass N.° 2.

g', g

Thousandths of an Inch

> **NUMEROUS BACTERIA, KNOWN AS *LISTERIA*, ARE NAMED AFTER LISTER, AND THE ANTISEPTIC MOUTHWASH LISTERINE IS NAMED IN HIS HONOUR.**

Like Semmelweis, Lister endured a great deal of early criticism. Colleagues elsewhere mocked his practices of wearing gloves for surgery and ensuring that both hands and instruments were washed in phenol solution. Part of the problem was phenol's irritating effects on skin, eyes, and lungs. Moreover, Lister was not a particularly effective advocate for his views. Eventually, however, the germ theory gained support and Lister was hailed as a hero.

After the death of his wife in 1893, Lister retired from practice, though he continued to serve as surgeon to the royal family for years. When King Edward developed appendicitis in 1902, Lister was on hand to ensure that the surgery was carried out with the most meticulous attention to antiseptic technique. Once the king recovered, he credited Lister with saving his life.

LEGACY

Lister's legacy is rich. He served as president of the Royal Society for five years. He is one of only two British surgeons with a public monument. Numerous bacteria, known as *Listeria*, are named after him, and the antiseptic mouthwash Listerine, developed by a St Louis chemist in 1879, is named in his honour. Today, the Royal College of Surgery's Lister Medal, originally awarded every three years (now every five), is widely regarded as the field's most prestigious mark of recognition.

Opposite: Sketches from Lister's microscopic observations.
Below: A gathering of Lister and his surgical colleagues.

19 FLORENCE NIGHTINGALE, THE "LADY WITH THE LAMP"

Florence Nightingale (1820–1910) contributed to the fight against infectious disease in several ways, including improved nutrition and sanitation during the Crimean War, the introduction of scientific education and practice for nursing, and the use of advanced statistical techniques to represent rates of infectious disease.

The daughter of wealthy British parents, Nightingale was named after the Italian city of her birth. Growing up on her family's estates, she experienced what she believed to be a divine call to devote her life to the service to others. Despite her family's strong objections, she resolved to pursue a career in nursing, a profession in poor repute at the time.

THE TRANSFORMATION OF NURSING

In 1854, Florence led a group of nurses to the Ottoman Empire, site of the Crimean War, where she was appalled by poor hygiene and high rates of death from what we now know to be infectious disease. Ten times as many soldiers died from infections as from battle wounds. Nightingale set to work to improve hygiene, including improved sewers and the introduction of handwashing.

Because of her habit of conducting rounds at night, Nightingale became known as the "lady with the lamp". Her reforms helped to reduce death rates sharply. Upon her return to England, in 1859 she wrote *Notes on Nursing*, and in 1860 she founded the Nightingale School at St Thomas' Hospital. From that point forward, she devoted her life to improving and promoting the profession of nursing.

Florence explained the nurse's role in these terms:

I use the word nursing for want of a better. It has been limited to signify little more than the administration of medicines and the application of poultices. It ought to signify the proper use of fresh air, light, warmth, cleanliness, quiet, and the proper selection and administration of diet – all at the least expense of vital power to the patient.

Nightingale was no proponent of the germ theory, but she believed that improvements in nutrition, hygiene, ventilation, and sanitation

Above, from top: Florence Nightingale; the Nightingale jewel.
Opposite: Florence Nightingale making her rounds by night.

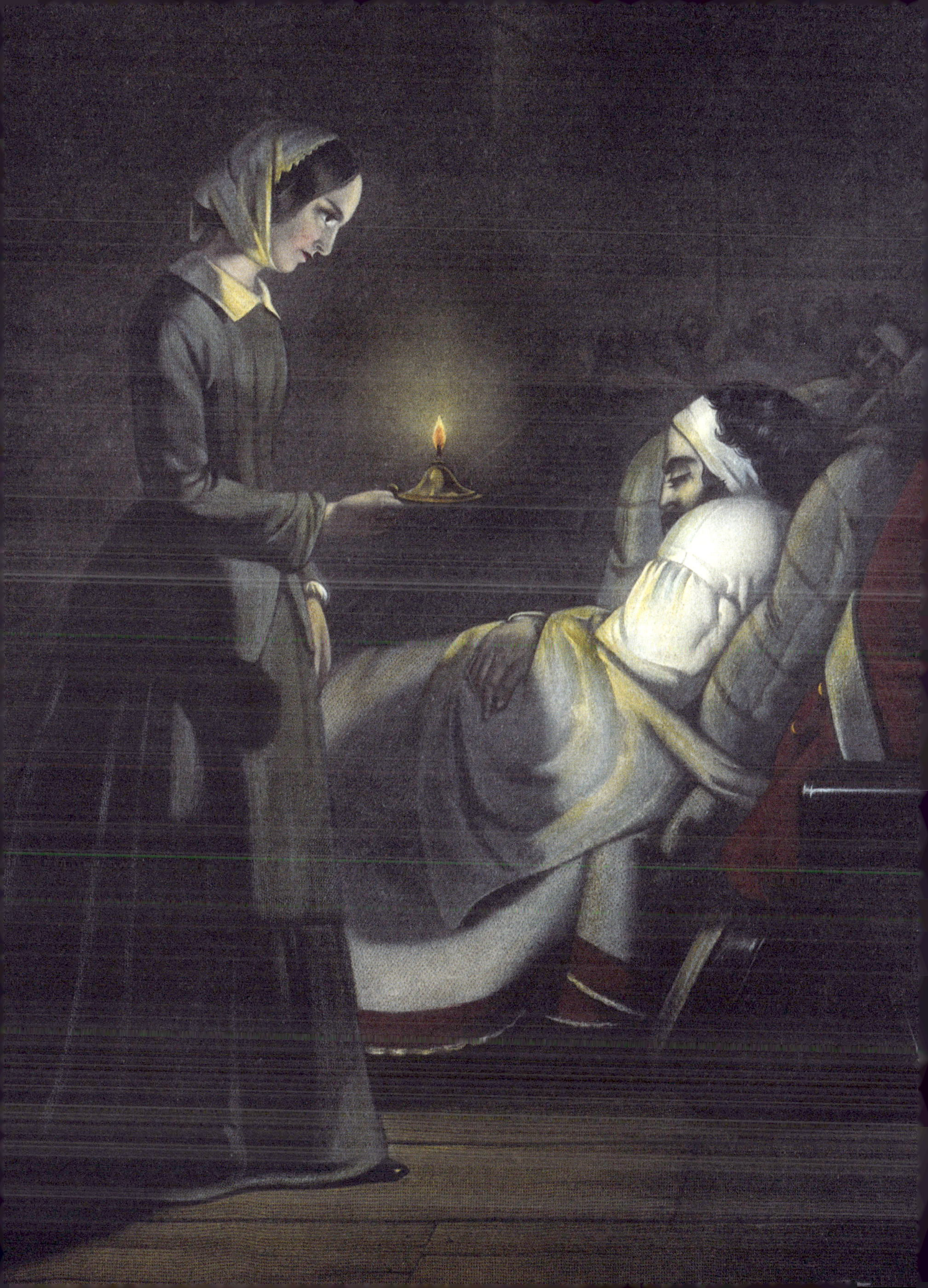

Letter from Miss FLORENCE NIGHTINGALE.

Dec 16/96
10, SOUTH STREET,
PARK LANE W.

Dear Duke of Westminster
Good speed to your noble effort in favour of District Nurses for town & country; and in Commemoration of our Queen who cares for all.
We look upon the District Nurse, if she is what she should be, & if we give her the training she should have, as the Great civilizer of the poor. training as well as nursing them out of ill health into good health (Health Missioners), out of drink into self control — but all without preaching, without patronizing — as friends in sympathy.
But let them hold the standard high as Nurses. Pray be sure, I will try to help all I can, tho' that be small, here I will with your leave let you know.
Pray believe me your Grace's faithful servant
Florence Nightingale

---

were essential. She was mathematically gifted and made major contributions in the use of graphics to illustrate statistical information. For example, while she was in the Crimea, she used histograms (graphical displays of data using bars of different heights) to illustrate changing patterns of patient mortality.

She characterized her attitude toward life as a form of progressivism:

The progressive world is necessarily divided into two classes – those who take the best of what there is and enjoy it – and those who wish for something better and try to create it. Without these two classes the world would be badly off. They are the very conditions of progress, both the one and the other. Were there none who were discontented with what they have, the world would never reach anything better.

LEGACY

Today, Nightingale is memorialized in many ways. The Nightingale Medal, awarded by the Red Cross, is nursing's highest distinction. Inspired by medicine's Hippocratic Oath, the Nightingale Pledge is recited by nursing students at the conclusion of their training. Florence's statue towers above Waterloo Place in London, and her image appeared on the British ten-pound note for many years.

Above: A letter from Florence Nightingale.
Opposite, from top: Nightingale's graphical depiction of causes of death; bedside nursing.

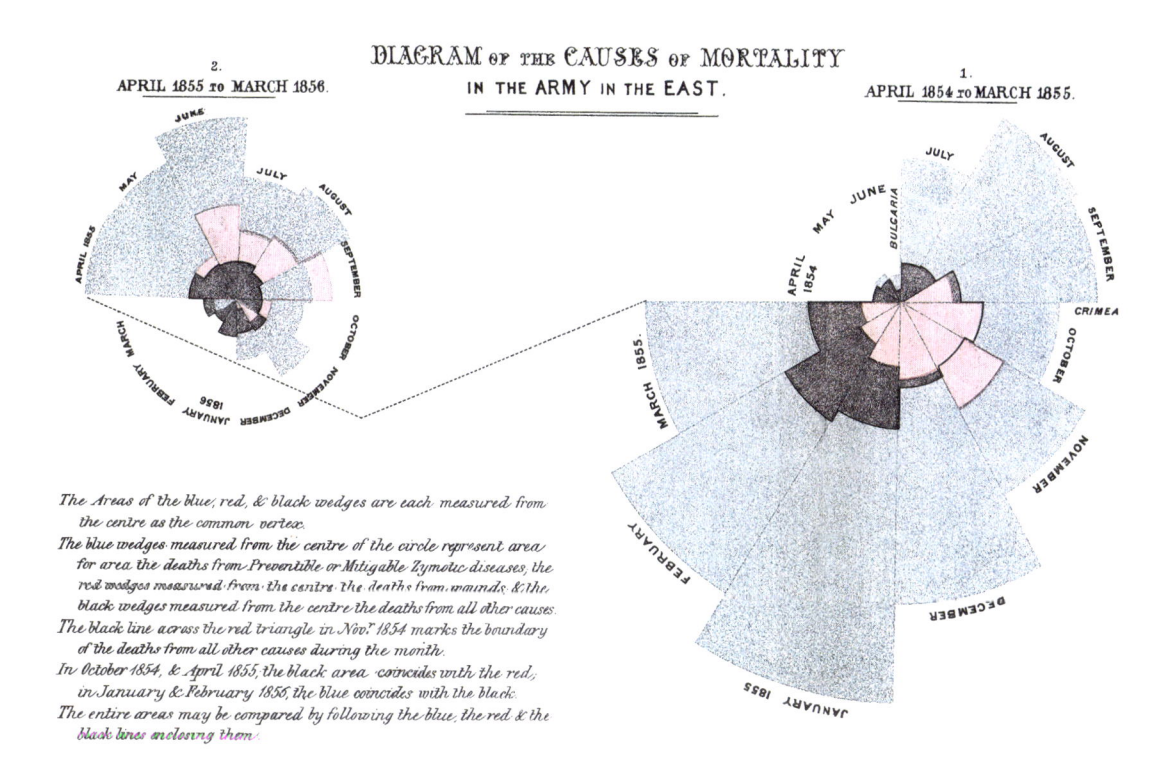

Florence Nightingale, the "Lady with the Lamp"

DIAGRAM OF THE CAUSES OF MORTALITY IN THE ARMY IN THE EAST.

2. APRIL 1855 TO MARCH 1856.

1. APRIL 1854 TO MARCH 1855.

The Areas of the blue, red, & black wedges are each measured from the centre as the common vertex.

The blue wedges measured from the centre of the circle represent area for area the deaths from Preventible or Mitigable Zymotic diseases, the red wedges measured from the centre the deaths from wounds, & the black wedges measured from the centre the deaths from all other causes.

The black line across the red triangle in Nov.r 1854 marks the boundary of the deaths from all other causes during the month.

In October 1854, & April 1855, the black area coincides with the red; in January & February 1856, the blue coincides with the black.

The entire areas may be compared by following the blue, the red & the black lines enclosing them.

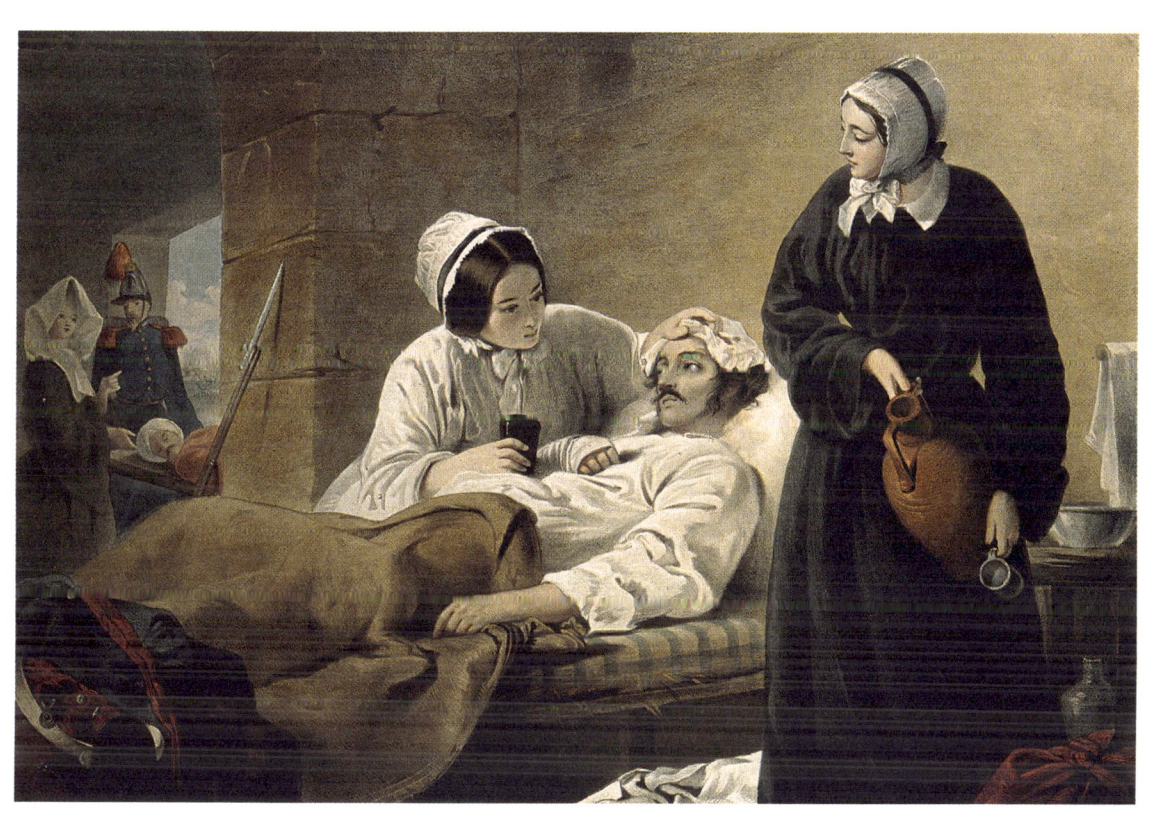

20 PASTEUR, MICROBIOLOGIST EXTRAORDINAIRE

The French chemist and microbiologist Louis Pasteur (1822–95) developed vaccines against two dreaded diseases, anthrax and rabies, but he also accomplished a great deal more: he established that biological molecules often exist in two mirror-image versions; that microbes are responsible for both fermentation of beer and wine and many diseases; and that the process of heating that came to be called pasteurization can prevent the growth of bacteria in beverages such as wine and milk.

Born in poverty in 1822, as a child Pasteur showed artistic talent. In school he studied philosophy, mathematics, and chemistry, in which he did not initially distinguish himself. With time, however, his performance improved, and he eventually took a position as a chemistry professor. In 1857, he moved to Paris, and in 1887, he established the Pasteur Institute, which he directed. He and his wife, Marie, had five children, but only two survived to adulthood, the other three dying from typhoid.

PASTEURIZATION

In 1857, Pasteur set out to demonstrate that yeast was responsible for transforming fruit juice into wine by turning sugar into alcohol. He also established that the contamination of wine, beer, and milk with microbes was responsible for their spoilage. He developed the process of pasteurization as a means of killing them. Inspired by this work, he also suggested that disease is caused when microbes gain entry to the human body, which spurred his investigation of antisepsis.

Having helped to secure the fortunes of the French wine-making and dairy industries by boosting hygiene, Pasteur was approached by members of the silk industry to study a disease afflicting silkworms. He developed a technique for examining moths to determine if they carried the disease; whenever signs of it were present, the eggs they laid were discarded. By culling out infected moths, silkworm farmers were able to dramatically reduce the disease's impact.

SPONTANEOUS GENERATION

Pasteur also became involved in the debate over spontaneous generation. He observed that when grapes were heated to a high temperature, their juice never fermented. This, and the fact that juice drawn by syringe from the centre of the grape also did not ferment, led him to conclude that yeast on the skin of the grape were responsible for fermentation – a view that contradicted that of other authorities, who argued that exposure to air alone caused the appearance of microbes.

Pasteur showed that when liquids were boiled in a flask, fermentation did not occur. He also used swan-neck flasks to show the same thing: although the contents were exposed to air, the long neck of the flask prevented microbial contamination. He then showed that by tilting the flask, exposing the liquid to the contaminated neck, fermentation did occur. He concluded that exposure to microbes in the air was necessary for fermentation, spoilage of meat, and so on.

Opposite: Pasteur at the laboratory bench.

Pasteur, Microbiologist Extraordinaire

PASTEUR AND ANTHRAX

Like the wine makers and silkworm farmers, cattlemen soon approached Pasteur for help, this time with the problem of anthrax. In response, Pasteur cultured the blood of anthrax-infected animals, then inoculated healthy animals, establishing that bacteria were responsible. When he subsequently discovered this bacteria in fields where farmers had buried diseased cattle, he warned them not to pasture their healthy animals in the same fields where they had buried ones that had died from anthrax.

Back in the lab, Pasteur discovered that by exposing anthrax bacteria to high temperatures, he could rob them of their ability to produce spores. A veterinarian proposed a test to see if inoculating healthy animals with the attenuated anthrax bacteria could render them immune to anthrax, an idea inspired by Edward Jenner's work with cowpox. The experiment proved successful, and Pasteur named the process vaccination (from the Latin *vacca*, for cow), in honour of Jenner's work.

Pasteur also proposed a vaccine for rabies, an almost universally fatal infectious disease of the nervous system now known to be caused by a virus. A nine-year-old boy had been bitten by a rabid dog, so at considerable personal risk, Pasteur administered a dozen doses of the vaccine that he believed contained an attenuated version of the microbe. Weeks later, the boy was still healthy, and Pasteur was hailed as a hero in the world press.

LEGACY

Pasteur became perhaps the most famous scientist in the world, receiving numerous international scientific awards. The rabies vaccine catalyzed the formation of the Pasteur Institute, with donations flowing in from around the world. The Institute, in turn, spawned the world's first course in microbiology. Today, there are Pasteur Institutes in dozens of countries around the world, and Pasteur Institute investigators won the 2008 Nobel Prize in Physiology or Medicine for identifying HIV, the retrovirus responsible for AIDS.

Pasteur was something of a philosopher as well as a scientist. Concerning scientific discovery, for example, he wrote that: "Fortune favors the prepared mind." Only those who are capable of recognizing what is

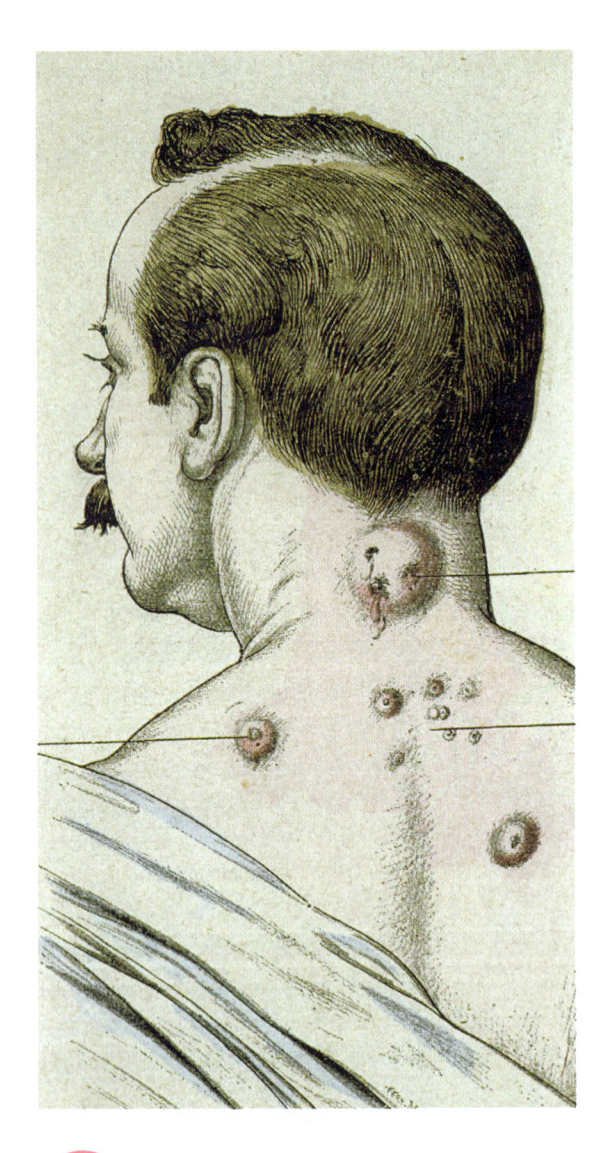

> **" ONE DOES NOT ASK OF ONE WHO SUFFERS, 'WHAT IS YOUR COUNTRY AND WHAT IS YOUR RELIGION?' ONE MERELY SAYS, 'YOU SUFFER, AND THAT IS ENOUGH FOR ME.'**
> **LOUIS PASTEUR**

truly intriguing and worthwhile will have the presence of mind to pursue leads that others might simply ignore. One goal of education, then, is to prepare the minds of learners to recognize when they have encountered something genuinely worthy of exploration.

Although questions have been raised about the ethics of Pasteur's work on the rabies vaccine, he promoted high ethical principles. For example, he held that a patient's national origin and faith should exert no influence on their access to care: "One does not ask of one who suffers, 'What is your country and what is your religion?' One merely says, 'You suffer, and that is enough for me.'" He expressed the belief that every person is equally human and equally deserving of care.

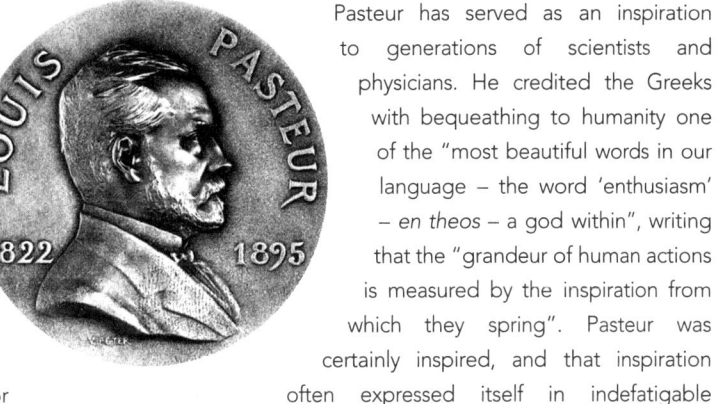

Pasteur has served as an inspiration to generations of scientists and physicians. He credited the Greeks with bequeathing to humanity one of the "most beautiful words in our language – the word 'enthusiasm' – en theos – a god within", writing that the "grandeur of human actions is measured by the inspiration from which they spring". Pasteur was certainly inspired, and that inspiration often expressed itself in indefatigable persistence. Indeed, as he once wrote, "my strength lies solely in my tenacity".

Opposite: Development of pustules in a patient suffering from anthrax.
Above: Medal with a bust of Pasteur.
Below: Microscope and tools used by Pasteur in his research on diseases of silkworms.

21 ROBERT KOCH AND HIS RADICAL POSTULATES

Pasteur and Koch are often regarded as the twin towers of nineteenth-century microbiology. Anyone who succeeded in isolating the causative organism of either anthrax, tuberculosis, or cholera would have been justly memorialized as a giant in the field, but Robert Koch (1843–1910) did so with all three. He also made key technological advances in cell culture and microscopy, and formulated his famous postulates for establishing the microbial causes of disease.

Koch was born the third of his parents' 13 children in northwest Germany. As a child, he proved a prodigy, learning to read before he arrived at school. Entering the University of Göttingen with the intention of studying science, he shifted to medicine, graduating with highest possible honours. He married the following year, fathering a single daughter.

Koch tended not to remain in one place for very long. He travelled around Germany, visiting many of its most prominent microbiologists. Receiving an appointment as a district health officer, he began investigating the disease anthrax. A disease of sheep and cattle, it can be transmitted to humans, causing skin ulcerations and a frequently fatal form of pneumonia.

KOCH AND ANTHRAX

Koch identified the disease in a variety of animal studies, and found that he could produce it by inoculating one species with infectious material obtained from another. At autopsy, he found rod-shaped bacteria in the infected creatures. He could then use them to inoculate another animal and reproduce the disease, again finding the rod-shaped bacteria.

At first, Koch did not actually know that the rod-shaped entities he was observing were bacteria, but he observed that they varied in length. The very long ones seemed to be dividing, leading him to conclude that he was

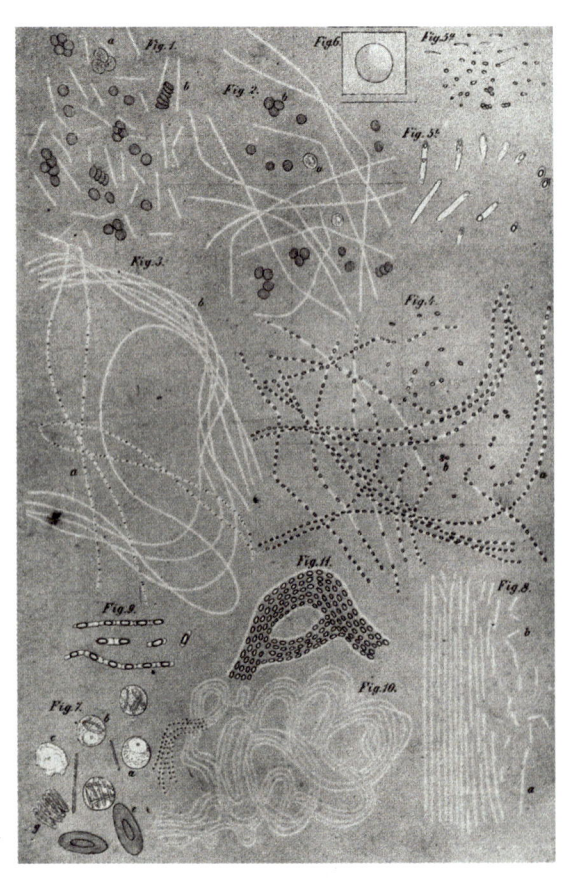

Above: Koch's depiction of anthrax bacilli.
Opposite: Cultures of different types of bacteria prepared by one of Koch's collaborators.

Robert Koch and his Radical Postulates

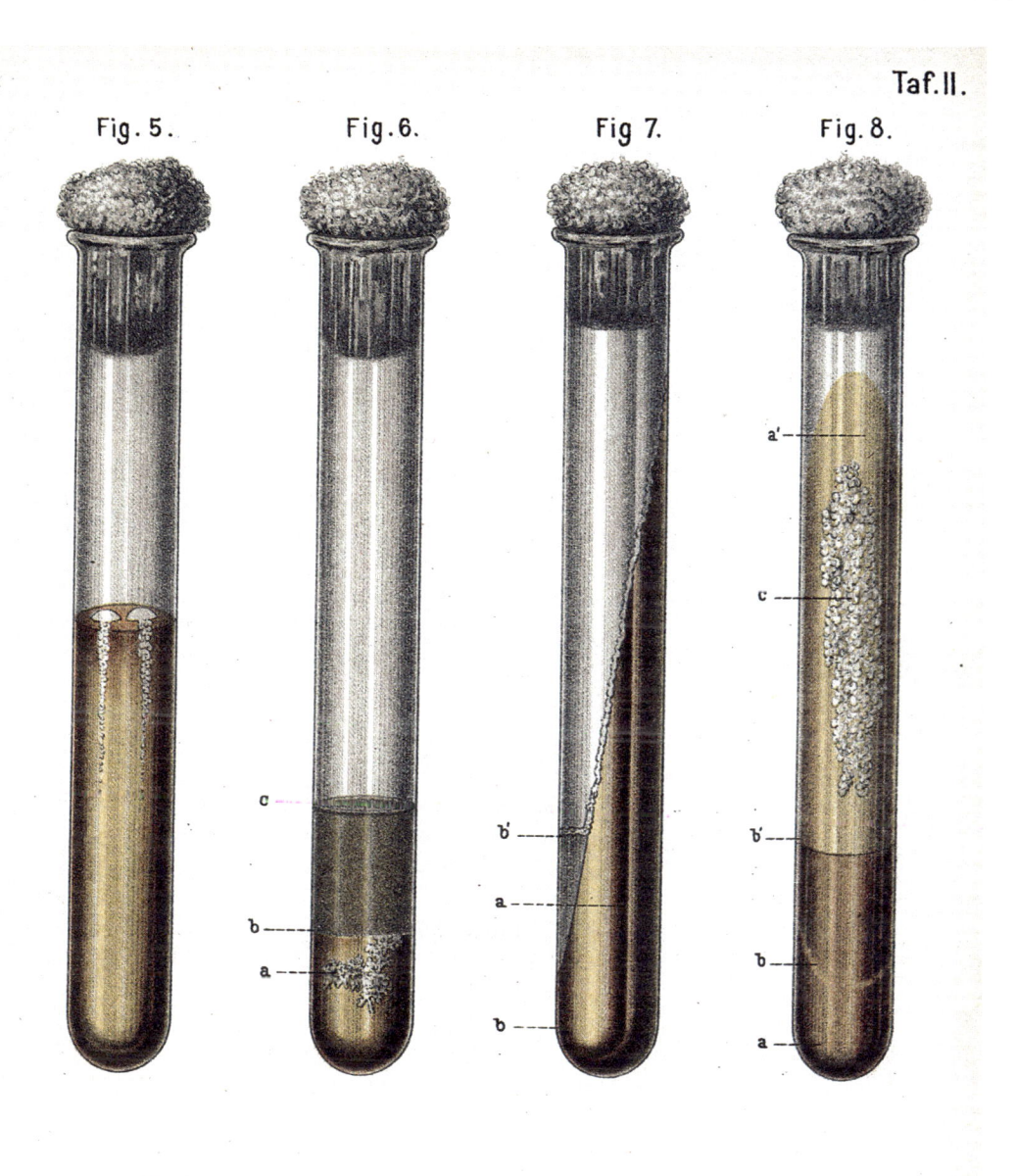

Taf. II.

Fig. 5.   Fig. 6.   Fig 7.   Fig. 8.

Fig. 9.   Fig. 10.   Fig. 11.

observing living organisms. Eventually, he learned how to grow them in culture, using the eye fluid of rabbits.

He demonstrated that the rod-shaped bacteria he was observing produced spores that persisted even after the bacteria had died. This explained how, though the bacteria could be short-lived, soil and other substances could remain infective for years. He recommended that animals that died from anthrax should be burned or buried in cold soil.

Koch's work on anthrax made him the first investigator to connect a specific disease with a specific microorganism and then to describe the life cycle of that organism – a major step forward for the germ theory of disease. Published in 1876, this work opened the floodgates to efforts of scientists around the world to connect specific infectious diseases with specific microbes.

KOCH'S CONTRIBUTION TO THE CULTIVATION OF PATHOGENS

We tend to think of Koch as a discoverer of pathogens, but his contributions to how pathogens are cultivated in the laboratory are equally notable. He sought to grow bacteria in pure culture, which he described as "the foundation for all research on infectious disease", an absolutely crucial step in isolating specific causative organisms. Building on discoveries by others, Koch found that agar, derived from seaweed, provided a good growth medium that could be placed in Petri dishes, named after his assistant.

This gave rise to the plate technique of bacterial culture. As opposed to a liquid medium, the plate could keep bacterial colonies separated, enabling him to subject them to a variety of toxins. Some, he found, would merely keep the colonies from growing (bacteriostatic), while others killed the bacteria outright (bactericidal).

KOCH AND TB

It was in connection with his research on tuberculosis that Koch formulated his famous postulates. He believed that the disease was not heredity, as many supposed, but was caused by a microorganism that resisted conventional stains used in microscopy. Experimenting with different compounds, he was able to develop stains that revealed previously invisible bacteria, including *Mycobacterium tuberculosis*.

Building on the work of colleagues, he formulated his postulates, which state that: the microbe must be present in diseased animals; the microbe must be isolated and grown in pure culture; when a disease-free animals is inoculated, it must develop the disease; and the microbe must again be isolated from it.

Today, we recognize numerous limitations of Koch's postulates. For example, some pathogens cause disease only in humans, resisting culture in animals; some, such as viruses, cannot grow in a cell-free culture; and some cause infectious disease only when present with other microbes. Yet Koch's postulates remain a kind of "gold standard" in microbiology to the present day.

Meticulously applying these postulates, Koch proved that a variety of diseases formerly thought to be separate, such as consumption, intestinal tuberculosis, and scrofula, were all manifestations of a single disease caused by the tubercle bacillus. He could take sputum from a tuberculosis patient and produce the disease in an animal.

Koch presented his findings in 1882, which propelled him almost immediately to worldwide fame. His work inspired many other pivotal figures in the subsequent history of microbiology, and in 1905 he received the Nobel Prize in Physiology or Medicine for his discovery. The discovery also paved the way for public health initiatives and laboratory research to contain the disease.

KOCH AND CHOLERA

Koch soon turned to the investigation of cholera. He led an expedition to Egypt and India to study the disease. Isolating a comma-shaped bacterium from the intestines of those who'd died of cholera, he found it difficult to reproduce the disease in animals, which showed the limitations of his own postulates. He was

> **HIS CREATIVITY, PERSEVERANCE, AND METICULOUS ATTENTION TO SOUND REASONING AND TECHNIQUE MAKE HIM ONE OF THE GREATEST FIGURES IN THE HISTORY OF MICROBIOLOGY.**

able to show, however, that no outbreaks occurred where the bacteria were absent.

Koch advised that the key to preventing this faecal–oral disease was the provision of clean drinking water. By filtering water supplies, municipal authorities could dramatically reduce rates of disease, or even eliminate it altogether. John Snow had shown that contaminated drinking water could cause disease, but Koch had shown what contamination really means.

LEGACY

Koch suffered some missteps along the way. For example, he expressed reservations about Pasteur's work, because the Frenchman was not a physician. His efforts to cure tuberculosis led him to develop tuberculin, an extract of the bacterium that he believed could halt the disease, but that proved not to be as effective as he hoped. He also mistakenly asserted that tuberculosis in cows did not cause human disease.

Reflecting on his work, Koch wrote:

If my efforts have led to greater success than usual, this is due, I believe, to the fact that during my wanderings in the field of medicine, I have strayed onto paths where the gold was still lying by the wayside. It takes a little luck to be able to distinguish gold from dross, but that is all.

Part of Koch's problem may have been his genius – he did not suffer individuals he regarded as sloppy or foolish gladly, and he frequently adopted a combative style in his writings and presentations. Nevertheless, his creativity, perseverance, and meticulous attention to sound reasoning and technique make him one of the greatest figures in the history of microbiology.

Opposite: Robert Koch.
Right, from top: Koch's depictions of bacteria, including those in tissue specimens; *Cholera bacilli*, as sketched by Koch.

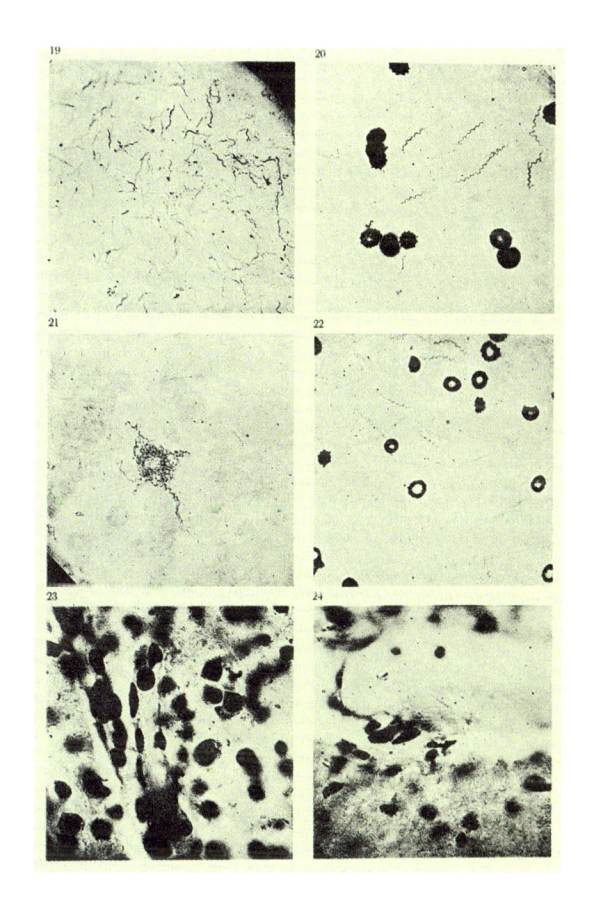

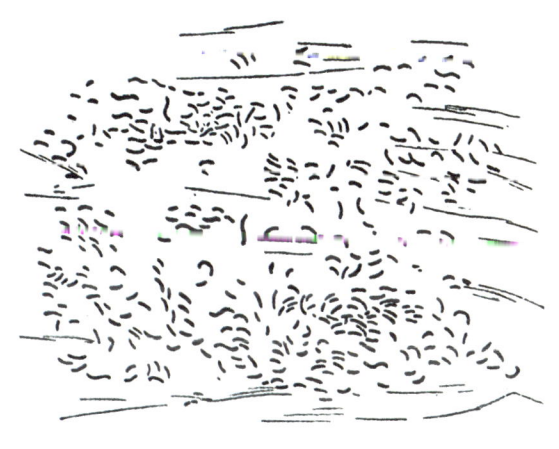

22 PETTENKOFER: GOOD EFFECTS FROM WRONG IDEAS

On 7 October 1892 German scientist Max Pettenkofer (1818–1901) took a drink. But it wasn't any ordinary draught. What Pettenkofer was imbibing was a dilute culture of *Vibrio cholerae*, then thought by some to be the causative agent of one of the most dreaded diseases of the nineteenth century: cholera. And Pettenkofer was not downing this dirty cocktail unwittingly. To the contrary, he knew exactly what he was doing.

A SCIENTIFIC RIVALRY

At the time, Pettenkofer was locked in a debate with Robert Koch, who believed that the bacillus his rival was blithely ingesting was the cause of a recent cholera epidemic in the city of Hamburg. Pettenkofer had determined that one way to prove Koch wrong would be to drink the culture Koch had provided and show that he did not develop the disease.

Some days after draining the draught, Pettenkofer pronounced these words of triumph:

Even if I had deceived myself and the experiment endangered my life, I would have looked death quietly in the eye, for mine would have been no foolish or cowardly suicide; I would have died in the service of science like a soldier on the field of honour. Health and life are very great earthly goods, but not the highest for man. Man, if we will rise above the animals, must sacrifice both life and health for higher ideals.

Ironically, the tide of history would turn against Pettenkofer. Although he had survived his perilous experiment, he did develop a short-lived diarrhoeal disease. And soon enough Koch and other proponents of the contagion school of disease would be vindicated. After the deaths of his wife and several children some years later, Pettenkofer himself, a great scientist and man of integrity, would be reduced to despair and take his own life.

BACKGROUND

Pettenkofer was born in Bavaria, now part of southern Germany. He studied medicine, graduating in 1845, and was then appointed a professor of chemistry, expanding in 1865 to a professorship in hygiene. His work as a chemist was sufficiently influential to be cited by Mendeleev as a source for his periodic table of the chemical elements. But it was through his work in hygiene that Pettenkofer made his greatest and most enduring contributions.

PETTENKOFER AND CHOLERA

Through meticulous research, Pettenkofer became convinced that inadequate sanitation was a major cause of human disease. Cholera had swept across Europe multiple times over the years, claiming the lives of substantial numbers of patients. Some experts argued that it was a communicable disease, because its spread seemed to follow trade routes. Pettenkofer disputed the idea that contagion was responsible, and he had good reasons for doing so.

Even as epidemics spread through a region, Pettenkofer argued, many areas remain disease free. The disease, he thus claimed, was the result of moist soil seeded with sewage, which helped to explain why crowded cities such as Lyon in France seemed to be spared. Lyon rested on a foundation of granite,

Above: Max Pettenkofer.
Right: manhole for access to the sewer system in Munich, which was designed by Pettenkofer.

which prevented the disease-causing decay that took place in the soil underlying other cities where epidemics developed.

This idea differed markedly from the theories of men such as Koch, who argued that cholera spread through the contamination of food or water by bacteria that multiplied in the human gut and were then transmitted by a faecal–oral route. Koch had found the organism responsible and satisfied himself that inoculation by the bacterium, transmitted from person to person, was the crucial step in the development of the disease, which could become an epidemic when water supplies were contaminated.

LEGACY AND LESSONS

Driven by his idea that cholera was bred in the soil, Pettenkofer made a number of salutary recommendations. The key, he felt, was to ensure that the soil remained dry and uncontaminated. As a result, he sought to bring clean water into cities and to drain away waste through sewage systems. Even though, today, most would say that Koch was right and Pettenkofer wrong, it is quite likely that Pettenkofer's recommendations saved hundreds of thousands and perhaps millions of lives.

For a number of reasons, Pettenkofer wielded considerable influence. He was widely regarded as a careful investigator and man of high character. His recommendations did lead to improvements in health. And perhaps most importantly, his analysis was favourable to economic and political leaders of the day, since they required no restrictions on international trade or domestic commerce to halt the spread of disease. By comparison, Koch faced an uphill battle.

As Pettenkofer approached the end of his life, the sympathies of the scientific community shifted increasingly toward Koch and those who believed that many diseases are contagious. Seeing the writing on the wall, in 1894 Pettenkofer retired from active research, although he continued writing, including a treatise on water purification that he completed in 1899. In despair over the fate of his life's work and the recent loss of his wife and several of their adult children, he shot himself in 1901.

Pettenkofer's story, though tragic, is in several respects an important one. It reminds us that even great minds looking at the same evidence can come to radically different conclusions. It also makes clear that a scientist need not be theoretically accurate to provide great benefit. Finally, and possibly most significantly, Pettenkofer's tale serves as a memorable example of the fact that the history of infectious disease is characterized by missteps and blind alleys that do not always lead directly to truth.

23

THE "GREATEST PANDEMIC IN HISTORY"

TEN THINGS WE ALL THOUGHT WE KNEW ABOUT THE SPANISH FLU

The year 2018 marked the centenary of the great influenza pandemic of 1918, which between January of that year and the December of 1920 is estimated to have infected half a billion people around the globe. Between 50 and 100 million are thought to have died, representing as much as 5 per cent of the world's population. Especially remarkable was the 1918 flu's predilection for taking the lives of otherwise healthy young adults, as opposed to children and the elderly, who usually suffer most. Some have called it the greatest pandemic in history.

The 1918 pandemic has been a regular subject of study and speculation over the past 100-plus years. Numerous hypotheses and conjectures have been advanced regarding its origin, spread, and consequences. As a result, many misconceptions have evolved and persisted about this devastating disease. By considering and correcting 10 of these misconceptions, we can arrive at a deeper and more accurate understanding of what actually happened, and derive lessons for helping to prevent and mitigate such disasters in the future.

1. THE PANDEMIC ORIGINATED IN SPAIN

No one believes the so-called "Spanish flu" originated in Spain. The pandemic probably acquired this nickname because of the First World War, which was in full swing at the time. The major countries involved in the war were naturally keen to avoid offering any encouragement to their enemies, so reports of the extent of the flu were suppressed in Germany, Austria, France, the United Kingdom, and the US. By contrast, neutral Spain had no need to keep the flu under wraps, which created the false impression that it was bearing the brunt of the disease. In fact, the geographic origin of the flu is unknown to this day, though hypotheses include East Asia, Europe, and even the US state of Kansas.

2. THE PANDEMIC WAS THE RESULT OF A "SUPER-VIRUS"

The rapidity of the 1918 flu's spread, which enabled it to kill 25 million people in just the first six months of the pandemic, led some to fear the end of humankind, and has long fuelled the supposition that the strain of influenza was particularly lethal. In fact, however, more recent study suggests that the virus itself was not a great deal more aggressive than other strains, and that much of the 1918 flu's high death rate can be attributed to crowding in military camps and urban environments, as well as poor nutrition and sanitation, associated at least in part with the war. It is now thought that many deaths were due to the development of secondary bacterial pneumonias.

Opposite: The influenza virus responsible for the 1918 flu pandemic.

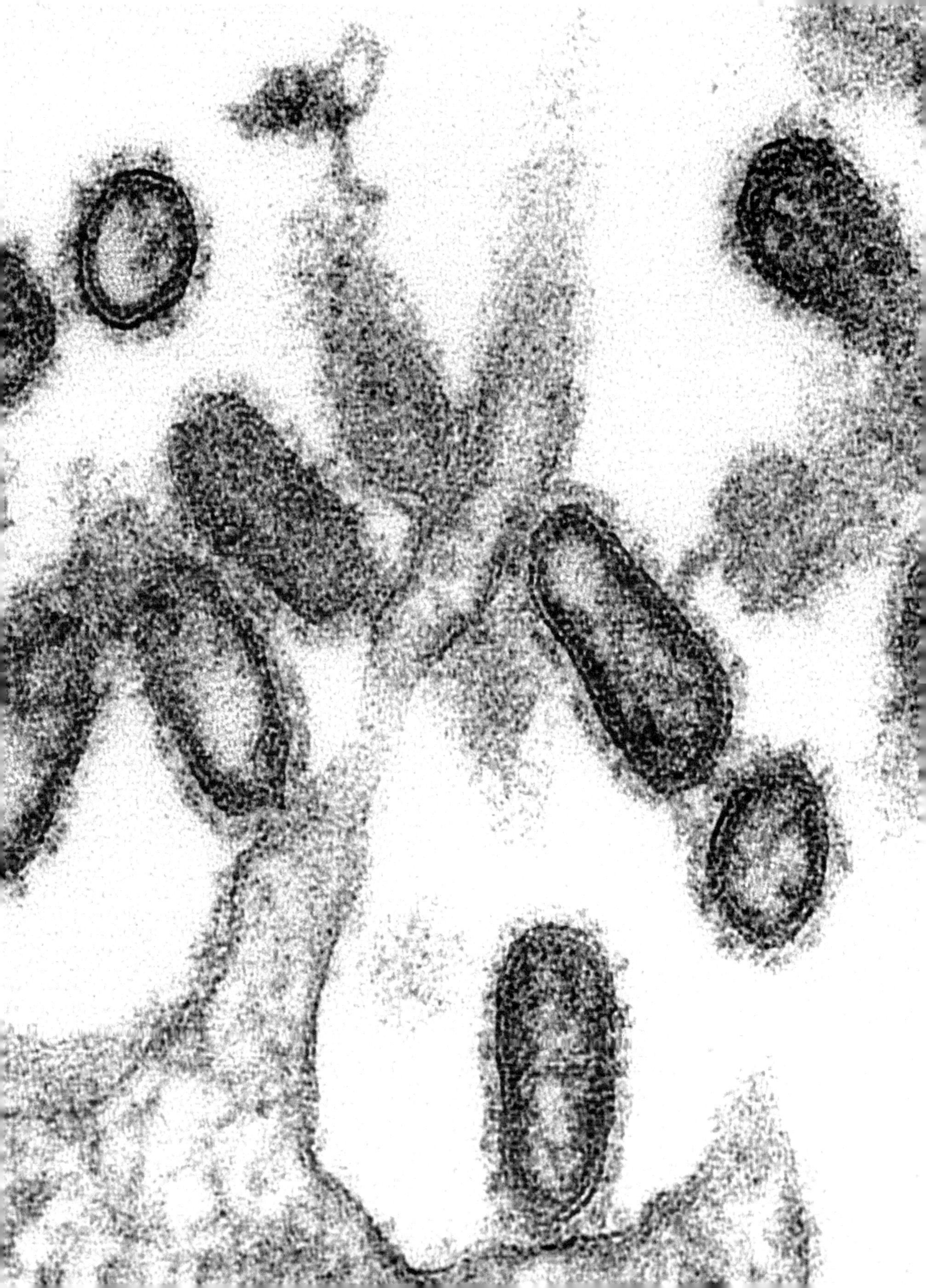

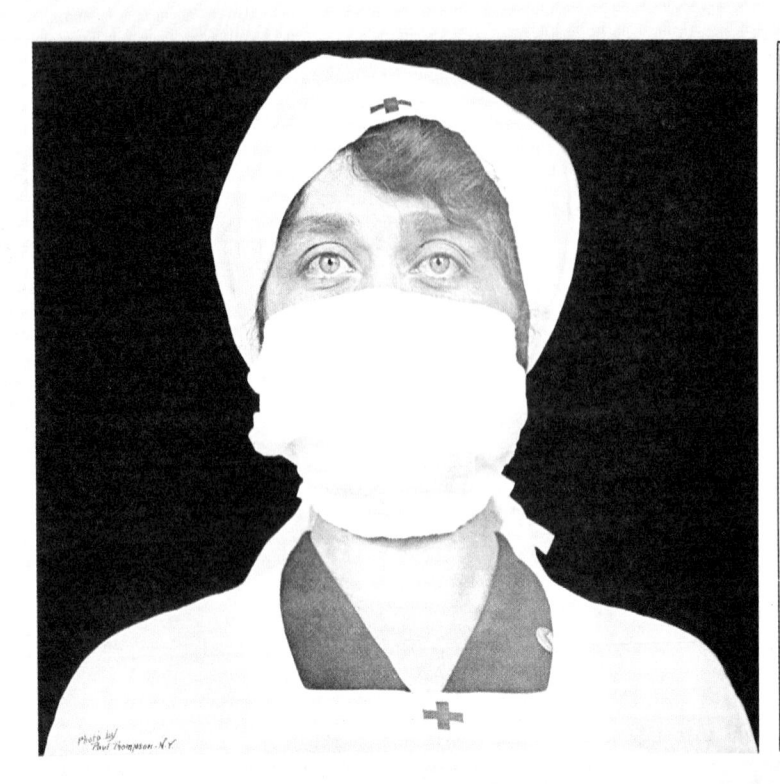

To Prevent Influenza!

Do not take any person's breath.
Keep the mouth and teeth clean.
Avoid those that cough and sneeze.
Don't visit poorly ventilated places.
Keep warm, get fresh air and sunshine.
Don't use common drinking cups, towels, etc.
Cover your mouth when you cough and sneeze.
Avoid Worry, Fear and Fatigue.
Stay at home if you have a cold.
Walk to your work or office.
In sick rooms wear a gauze mask like in illustration.

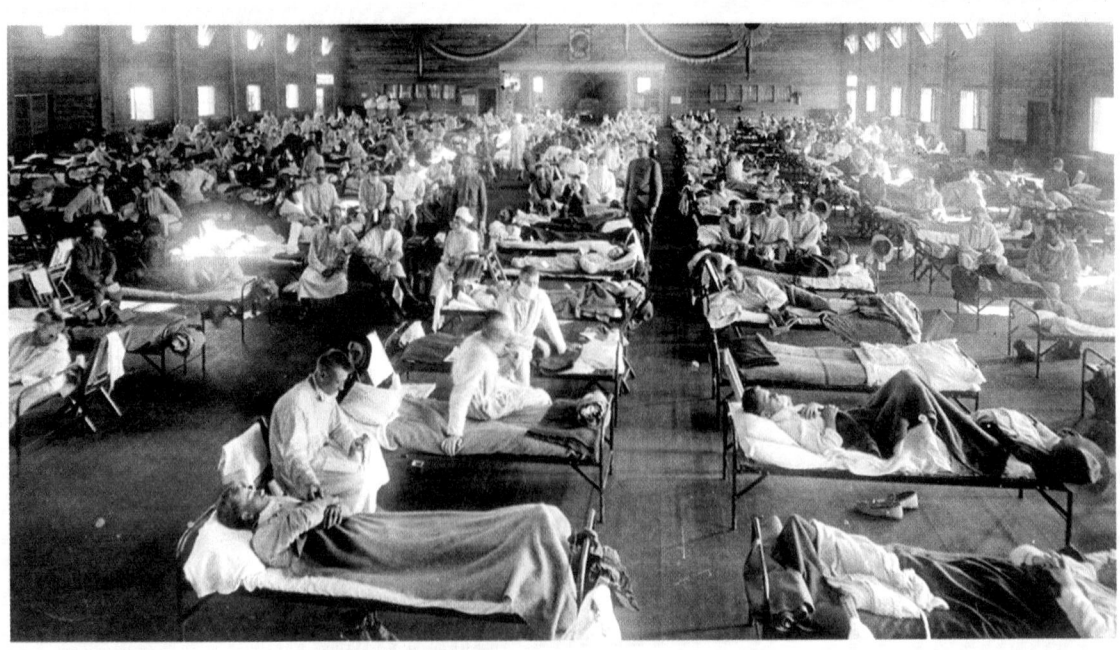

Above, from top: A picture of a masked Red Cross nurse is used in this public information advertisement designed to slow down the spread of the virus; Camp Funston, Kansas. This is the ward where the epidemic first made its major appearance among First World War veterans in 1918.

Left: Jeffery Taubenberger, virologist at the US National Institute of Allergy and Infectious Diseases.

3. THE FIRST WAVE OF THE PANDEMIC WAS MOST LETHAL

To the contrary, the first wave of deaths from the pandemic in the first half of 1918 demonstrated a relatively low rate of lethality, and it was during the second wave from October through December of that year that the highest death rates were observed. A third wave in the spring of 1919 was more lethal than the first but less so than the second. The markedly increased death rates of the second wave are thought to have been due to mutation of the virus to a deadlier strain. Changes in the virus were likely amplified by human responses to it. Those with mild cases stayed at home, but those with severe cases were often crowded together in hospitals and camps, increasing transmission of the deadlier strain.

4. THE VIRUS KILLED MOST PEOPLE WHO WERE INFECTED WITH IT

In fact, the vast majority of the people who contracted the 1918 flu survived, and national death rates among the infected generally did not exceed 20 per cent. However, mortality varied among different groups, and in the US, it was particularly high among Native American populations, among whom in some cases entire communities were wiped out. The symptoms of the disease were unusual, and in some cases included haemorrhaging from the nose, ears, and intestines, leading some public health experts to think that they were dealing with another disease entirely, such as dengue or typhoid. Of course, even a 20 per cent death rate vastly exceeds the typical flu mortality of less than 1 per cent.

5. THERAPIES OF THE DAY HAD LITTLE IMPACT ON THE DISEASE

It is true that no specific antiviral therapies were available during the 1918 pandemic, a situation that persists to a substantial degree today, when supportive care is still the rule. However, one investigator has proposed that many flu deaths were due to aspirin poisoning. Supporting this hypothesis is the fact that medical authorities recommended large doses of aspirin of up to 30 g per day (today about 4 g would be considered the maximum safe daily dose) – large doses of aspirin can lead to many of the supposed symptoms noted during the pandemic, including bleeding. However, the death rates seem to have been equally high in places in the world where aspirin was not so readily available.

6. THE PANDEMIC DOMINATED THE DAY'S NEWS

Public health officials, law enforcement officers, and politicians had reasons to underplay the severity of the 1918 flu. As already noted, there was fear that full disclosure might discourage allies and embolden enemies during wartime. Another factor was the desire to preserve public order and avoid panic. However, officials did respond. At the height of the pandemic, quarantines were instituted in many cities, and some even shut down essential services, including police and fire. Of course, many individuals and communities also responded with heroism, and the contributions of nurses during the pandemic did much to heighten public regard for the profession.

7. THE PANDEMIC CHANGED THE COURSE OF THE FIRST WORLD WAR

It is unlikely that the flu of 1918 changed the outcome of the First World War, because combatants on both sides of the battlefield were relatively equally affected. However, there is little doubt that the war profoundly influenced the course of the pandemic. For one thing, the mobilization and concentration of millions of troops provided ideal circumstances for the development of more virulent strains of the virus and its spread around the globe. The fact that soldiers who had served in the military for years suffered lower rates of death than new recruits supports the hypothesis that exposure to prior strains of the flu offered some protection.

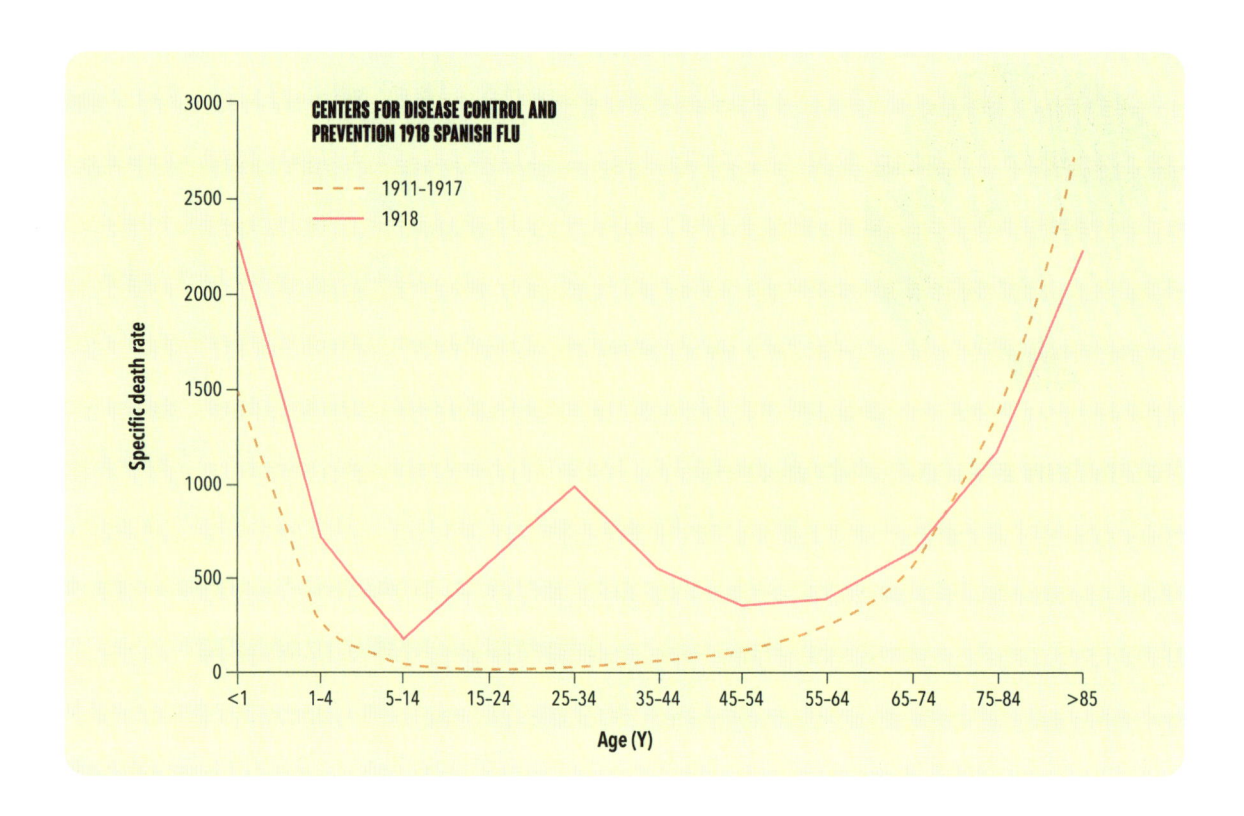

8. WIDESPREAD IMMUNIZATION ENDED THE PANDEMIC

In fact, immunization against influenza as we know it today was not practised in 1918 and thus played no role in ending the pandemic. On the other hand, the development of immunity through the natural course of infection played a pivotal role. For example, those infected during the relatively benign first wave of the disease were protected from infection during the highly lethal second wave. Another factor in the end of the pandemic was probably that the rapidly mutating virus developed into less lethal strains. This is predicted by models of natural selection, which show that highly lethal strains have less opportunity to spread (the host dies rapidly) than less lethal strains, thus favouring the latter.

9. THE GENES OF THE VIRUS HAVE NEVER BEEN SEQUENCED

In 2005, researchers announced that they had successfully determined the gene sequence of the 1918 influenza virus. The virus was recovered from the body of a flu victim who had been buried in the permafrost of Alaska, as well as from samples of American soldiers who fell ill at the time. Two years later, monkeys infected with the virus were found to exhibit the symptoms observed during the pandemic. Pathologic analysis suggested that the monkeys died from an overreaction of the immune system to the virus, the so-called cytokine storm – which helps to explain why the death rates were highest among otherwise healthy young adults.

10. THE 1918 PANDEMIC OFFERS FEW LESSONS FOR TODAY

Influenza epidemics tend to occur every 30 to 40 years, and experts believe that the next one is a question not of "if" but "when". Among lessons learned from the 1918 flu pandemic that can be applied today are the importance of immunization, particularly of vulnerable populations; surveillance systems to detect the next epidemic at a more containable early stage; the use of antibiotics, not available in 1918, to combat secondary bacterial infections; and the proactive development of public health disaster plans, including isolation and the handling of large numbers of seriously ill and dying patients. Perhaps the best hope today lies in widely improved nutrition, sanitation, and standards of living.

The "Greatest Pandemic in History"

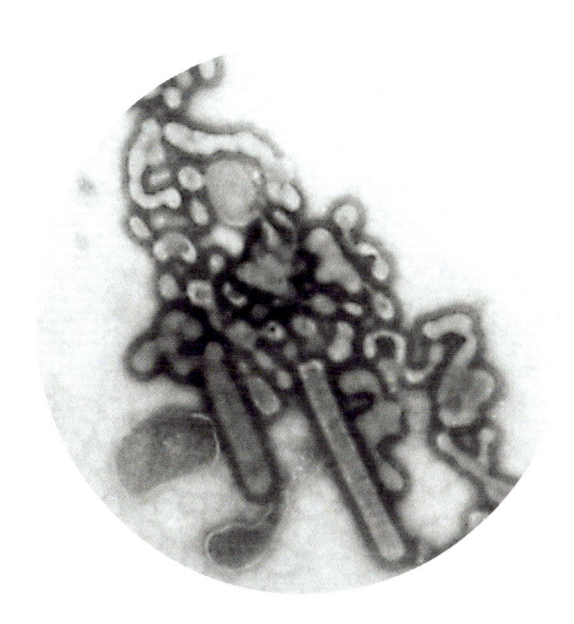

Above: Electronic micrograph of influenza A virus.
Right: A New York traffic policeman wearing a gauze mask at the height of the epidemic in October 1918.

FLU TODAY

Most people have suffered the fever, runny nose, sore throat, cough, muscle aches, headache, and fatigue that generally accompany the flu. While the great flu pandemic of 1918 has faded from living memory, we can continue to learn from its lessons, which range from the common-sense value of handwashing and immunizations to the great potential of cutting-edge research on influenza biology and the development of antiviral drugs. For the foreseeable future, flu epidemics will remain an annual feature of the rhythm of human life, and we can only hope that we have learned the great pandemic's lessons sufficiently well to mitigate another such worldwide catastrophe.

> **INFLUENZA EPIDEMICS TEND TO OCCUR EVERY 30 TO 40 YEARS, AND EXPERTS BELIEVE THAT THE NEXT ONE IS A QUESTION NOT OF "IF" BUT "WHEN".**

GLOBAL SPREAD OF THE SPANISH FLU
second wave, late 1918

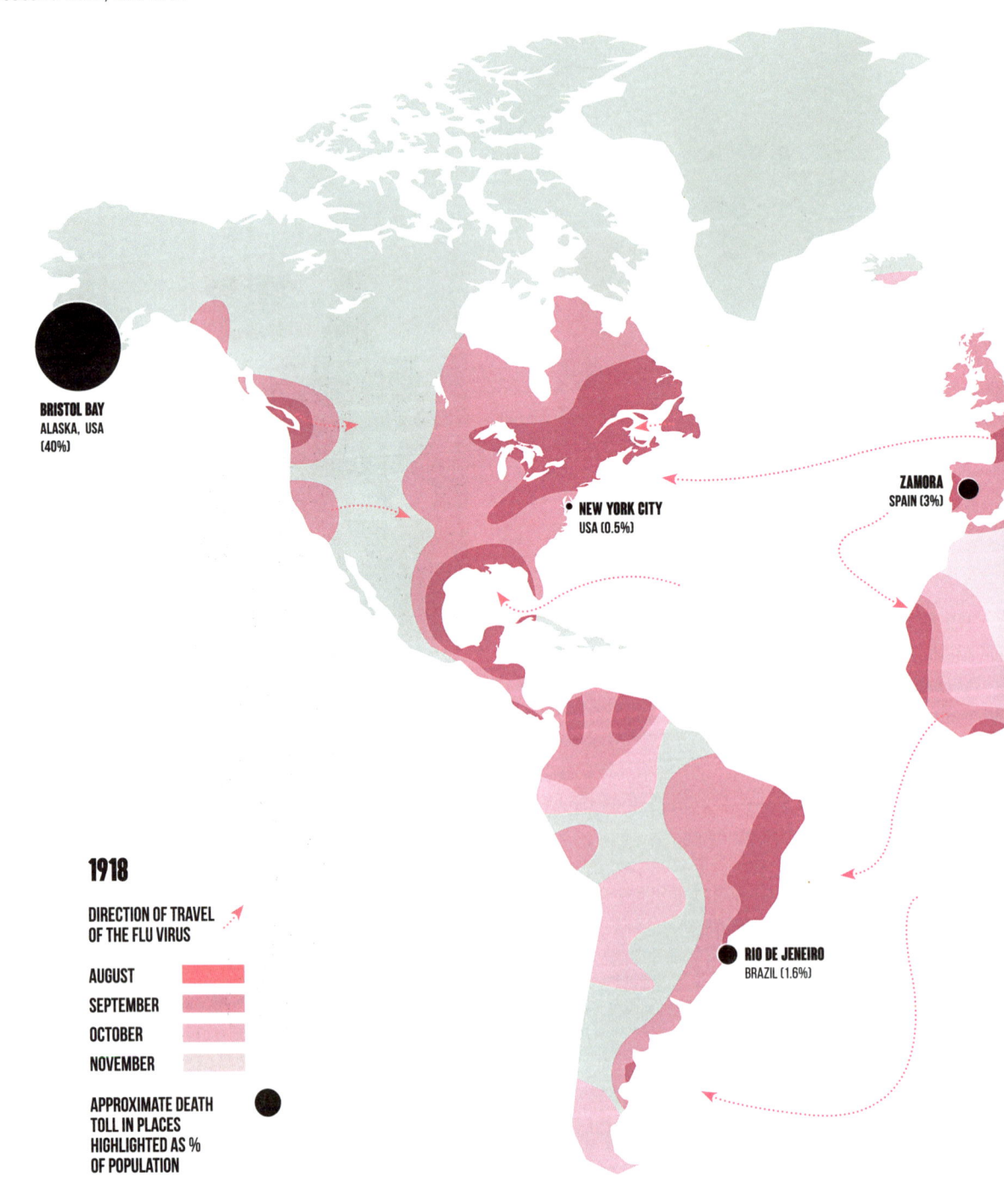

BRISTOL BAY
ALASKA, USA
(40%)

NEW YORK CITY
USA (0.5%)

ZAMORA
SPAIN (3%)

RIO DE JENEIRO
BRAZIL (1.6%)

1918

DIRECTION OF TRAVEL
OF THE FLU VIRUS

AUGUST
SEPTEMBER
OCTOBER
NOVEMBER

APPROXIMATE DEATH
TOLL IN PLACES
HIGHLIGHTED AS %
OF POPULATION

The "Greatest Pandemic in History"

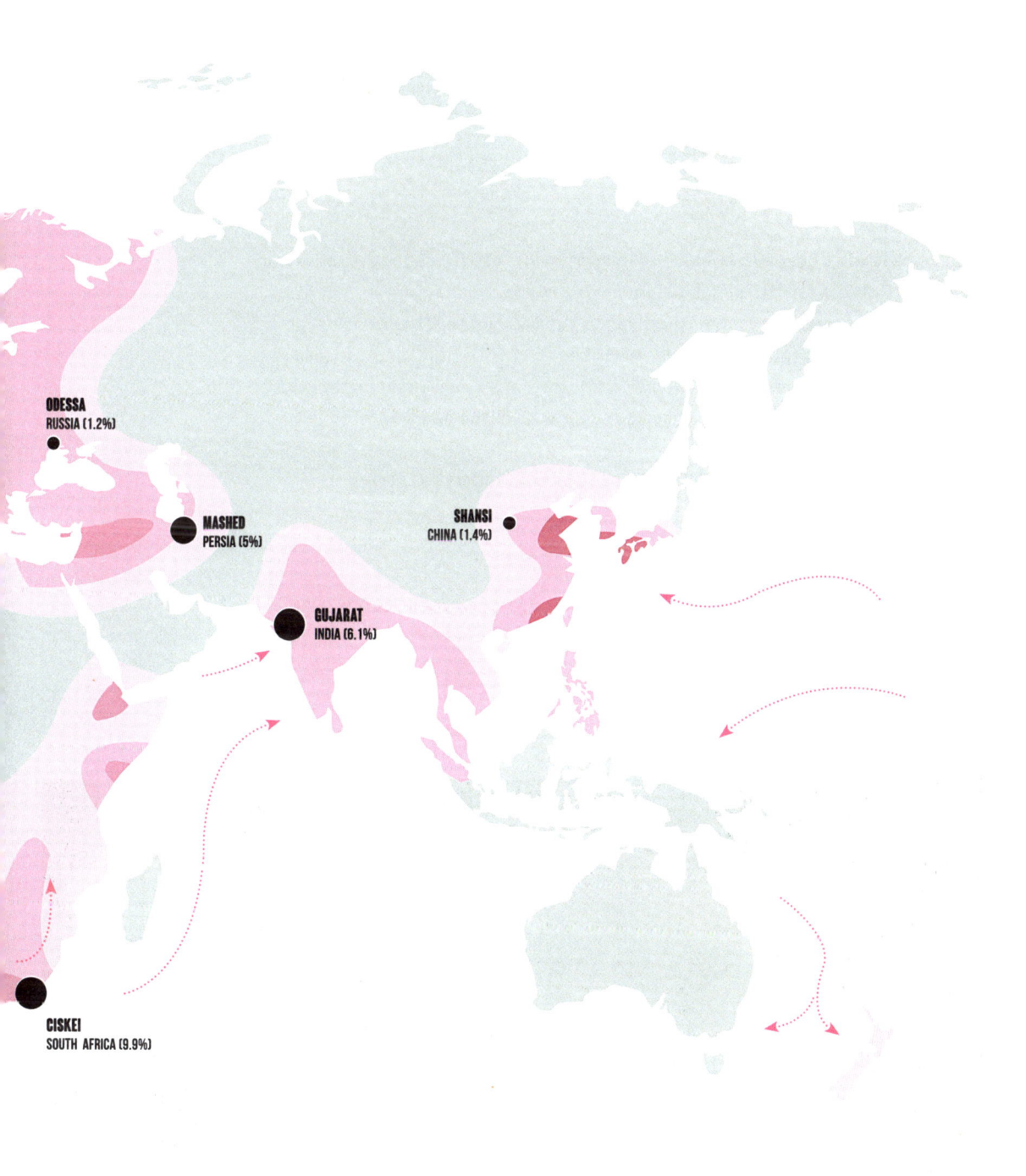

ODESSA
RUSSIA (1.2%)

MASHED
PERSIA (5%)

SHANSI
CHINA (1.4%)

GUJARAT
INDIA (6.1%)

CISKEI
SOUTH AFRICA (9.9%)

24 THE WORLD'S DEADLIEST ANIMAL

What is the world's deadliest animal? To answer such a question, it is first necessary to know what we mean by animal. An animal is a multicellular organism that must eat other living organisms or their products to survive, and that breathes oxygen, moves about, and reproduces sexually. Microbes such as bacteria and viruses are not animals. It is estimated that the earth harbours some 7 million animal species that may have arisen from a single ancestor over 600 million years ago.

Asked to name the deadliest animal, many people would think first of sharks, bears, or snakes, but they would be wrong. In fact, there is one very familiar large animal that kills more human beings than all these animals combined, and that animal, of course, is human beings. In a typical year, humans kill nearly a half million of their own kind, and in periods of wide conflict, that number increases even further. For example, 70 million people died during the Second World War.

SMALL BUT DEADLY

But outside of such conflicts, humans are not the principal culprit. The title belongs to a relatively small creature that comes in over three thousand different species. Its name means little fly, and these creatures lay their eggs on the surface of water, hatch into larvae, and then mature into winged insects. These mature creatures then feed on the blood of a variety of creatures, including amphibians, reptiles, bird, and mammals – most notably, human beings.

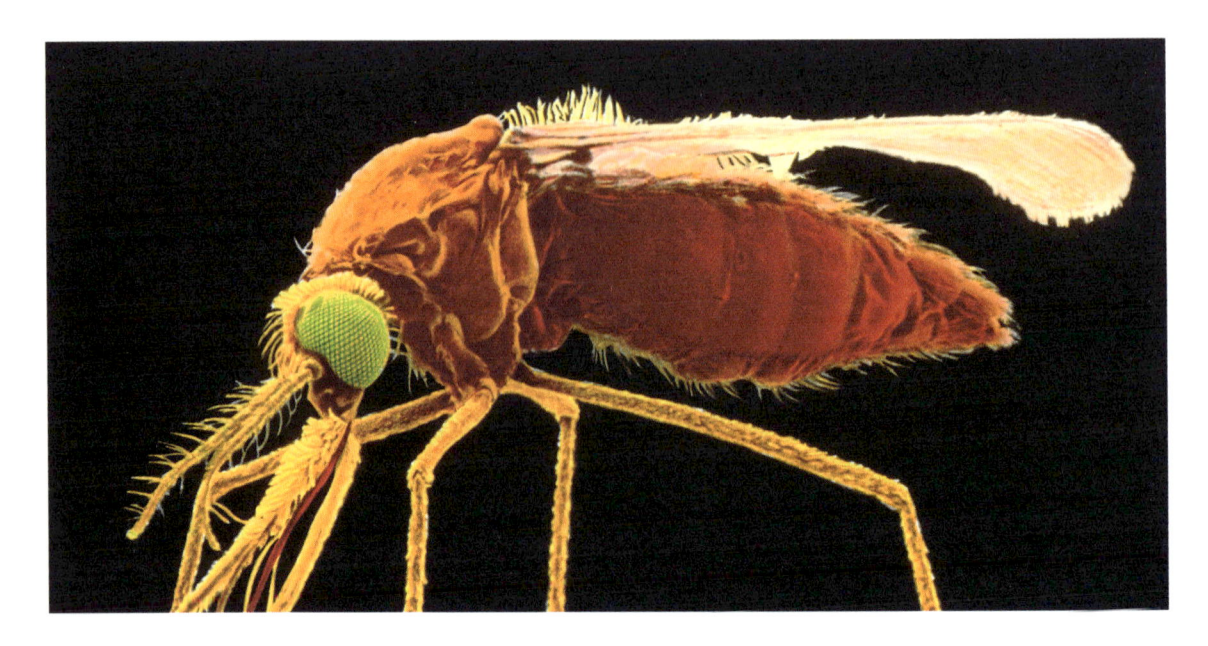

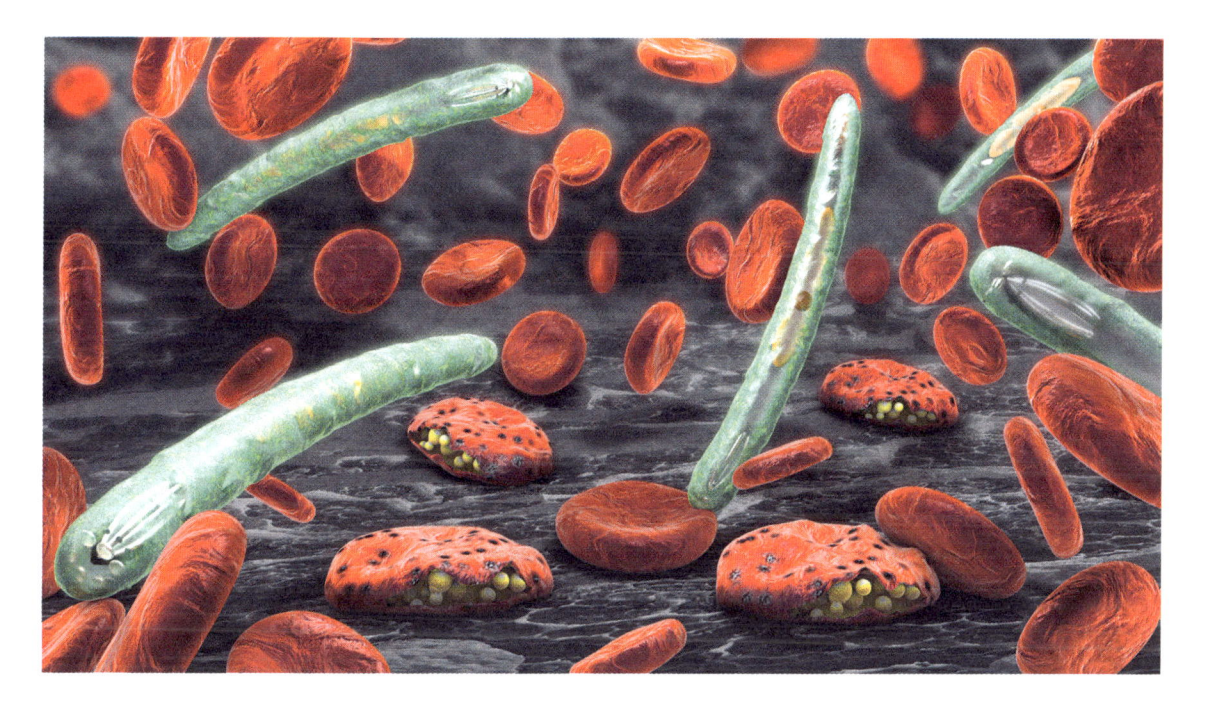

The world's deadliest animal, at least where human beings are concerned, is the mosquito. It is estimated that as many as 1 million people have died each year in recent decades from mosquitoes and the infections they transmit, most prominently malaria. Certain types of mosquitoes can carry a single-celled *Plasmodium* organism, which is introduced into the human bloodstream through the mosquito's saliva when a female seeks a blood meal in preparation for reproduction.

Malaria gets its name from "bad air", a problem associated by ancient physicians with swamps and marshlands. A French physician, Charles Laveran, received the 1907 Nobel Prize in Physiology or Medicine in part for discovering malaria parasites within red blood cells. Shortly thereafter, a Cuban physician, Carlos Finlay, gathered evidence that mosquitoes were involved in transmitting the disease to humans.

Evidence of malaria-causing organisms can be found in amber dating back tens of thousands of years. Today, because mosquitoes are usually active in warmer temperatures, malaria tends to be endemic in equatorial regions, and it is found mainly in Africa and Southeast Asia. Over 200 million new infections are thought to occur each year, with children accounting for about two-thirds of deaths. The numbers of new infections each year in Europe and the US is far lower, numbering in the thousands.

SYMPTOMS AND PROGRESSION

Symptoms usually begin about 10 to 20 days after infection, with a flu-like illness that includes headache, fever, and muscle aches. More severe symptoms include respiratory distress, seizures, and coma, and severe cases can progress to death. The classic symptoms of malaria are cyclical, with so-called paroxysms of fever occurring every two to three days, corresponding to the release of the microbe into the bloodstream.

The *Plasmodium* organism is a protozoon. Once injected into the bloodstream by a mosquito bite, the organisms migrate to the liver, where they reproduce without causing symptoms for a period of weeks. Once they escape the cells of the liver, they re-enter the bloodstream and begin infecting red blood cells. As they break out of some red blood cells and infect others, they cause the classic waves of fever.

Opposite: A mosquito.
Above: *Plasmodium* cells among red blood cells.

Man Made Malaria

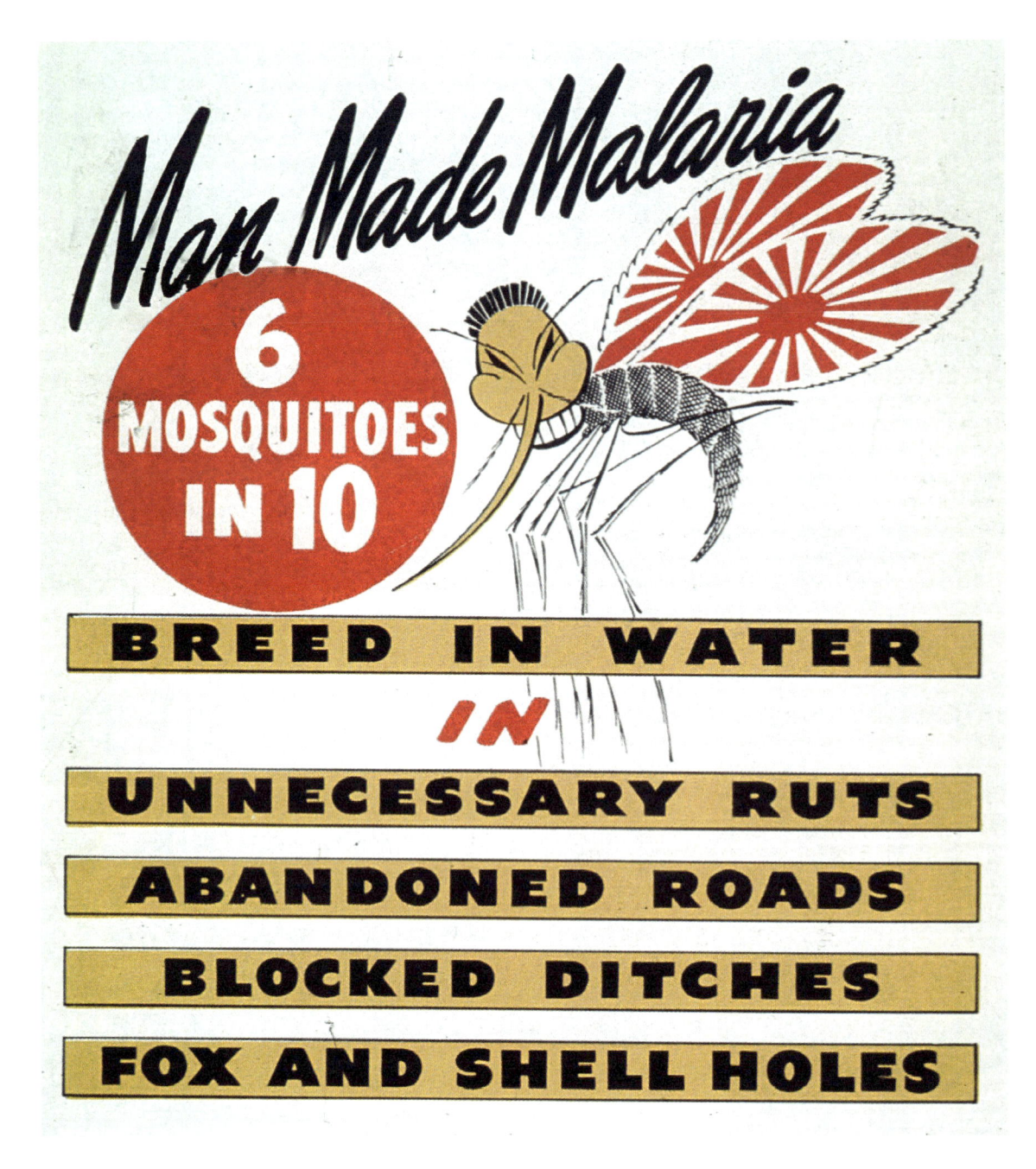

6 MOSQUITOES IN 10

BREED IN WATER IN
UNNECESSARY RUTS
ABANDONED ROADS
BLOCKED DITCHES
FOX AND SHELL HOLES

Malaria is difficult for the immune system to eradicate, in part because the *Plasmodium* parasite remains hidden within liver and red blood cells much of the time. However, infected red blood cells are recognized by the body as defective, and they are taken up and recycled in the spleen. The *Plasmodium* parasite can avoid this fate by attaching infected red blood cells to the walls of small blood vessels, which can interfere to some degree with oxygen delivery.

Unfortunately, effective vaccines against malaria have not yet been introduced, but there are means

Above: Poster warning against mosquito breeding grounds.

to combat the disease, including prevention of transmission and medications. For malaria to become endemic in an area, three factors are necessary: mosquitoes must be present in relatively large concentrations, as must human beings, and there must be relatively high rates of transmission from mosquitoes to human beings. This provides several ways to reduce or eliminate the disease.

COMBATING MALARIA

One means of combating malaria is to attack its vector: the mosquito. Insect repellents can keep mosquitoes from biting and insecticides can kill them. One insecticide that achieved wide use in the twentieth century was DDT. Paul Muller, who discovered its insecticidal properties, received the Nobel Prize in Physiology or Medicine in 1948. However, the 1962 publication of Rachel Carson's book *Silent Spring* helped draw attention to DDT's adverse environmental impact, especially on birds.

Another means of reducing malaria is the use of nets that interpose a physical barrier between mosquitoes and humans, particularly at night. To enhance their effectiveness, such nets can be treated with insect repellents or insecticides. In areas where malaria is endemic, such as parts of Africa, it is estimated that tens of millions of children have been protected by nets, although many more currently remain unprotected.

Unsurprisingly, when mosquito populations are treated with insecticides over long periods of time, they can develop resistance, due to the same forces of natural selection responsible for antibiotic resistance among bacteria. Other means of reducing mosquito populations include draining or covering collections

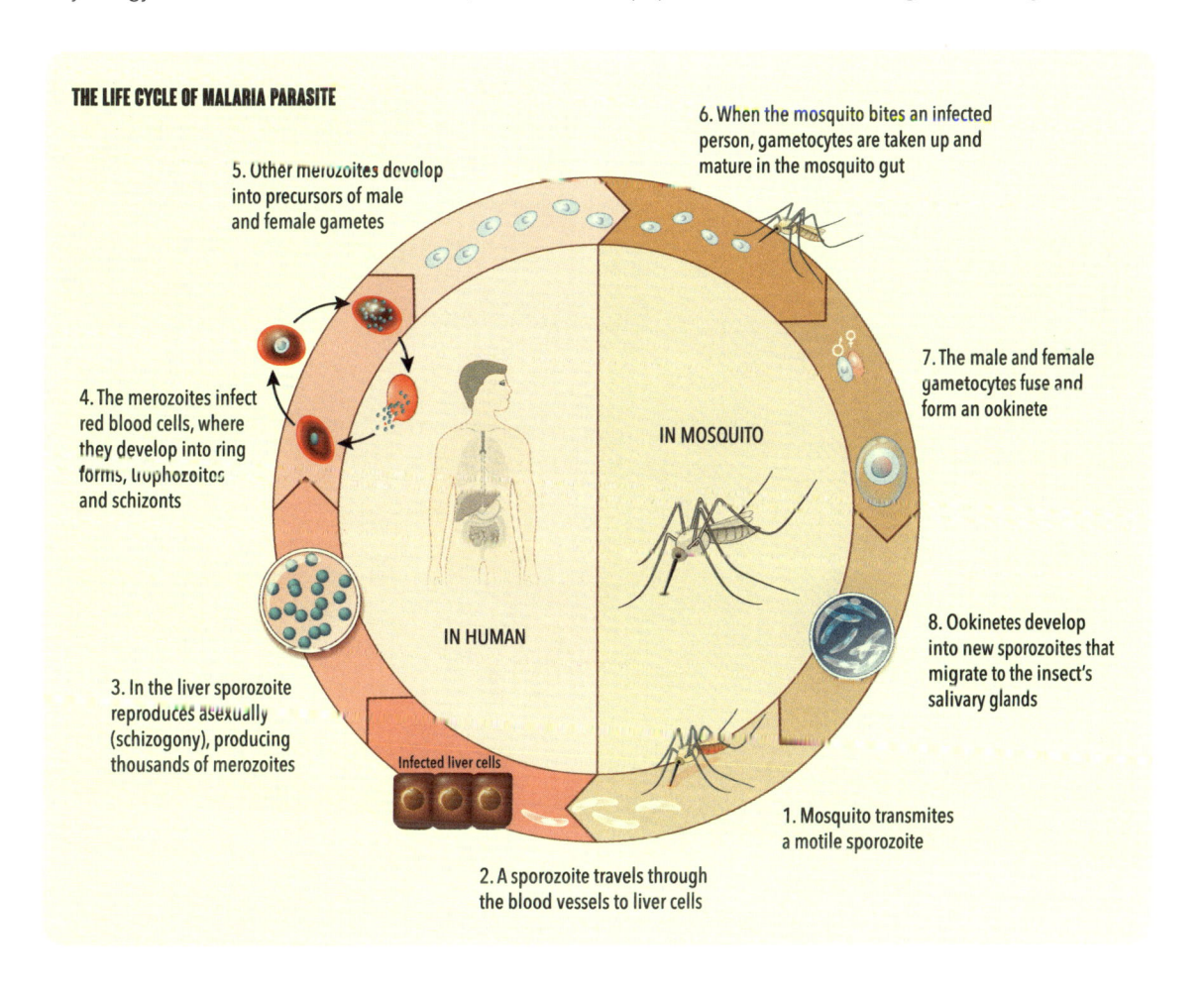

THE LIFE CYCLE OF MALARIA PARASITE

1. Mosquito transmites a motile sporozoite
2. A sporozoite travels through the blood vessels to liver cells
3. In the liver sporozoite reproduces asexually (schizogony), producing thousands of merozoites
4. The merozoites infect red blood cells, where they develop into ring forms, trophozoites and schizonts
5. Other merozoites develop into precursors of male and female gametes
6. When the mosquito bites an infected person, gametocytes are taken up and mature in the mosquito gut
7. The male and female gametocytes fuse and form an ookinete
8. Ookinetes develop into new sporozoites that migrate to the insect's salivary glands

of standing water where they breed, an initiative that can prove especially effective in areas of high human population density.

Malaria infections can be treated with oral medications, which are now often administered in combination, to decrease the chances of resistance developing to any single drug. A group of effective drugs called artemisinins, based on Chinese traditional medicines, was identified in the 1970s by Tu Youyou, who shared the 2015 Nobel Prize in Physiology or Medicine for her discovery. Other drugs long used against malaria include quinine and chloroquine.

Sources of hope in the battle against this killer disease include the development of new drugs against *Plasmodium* species, as well as the introduction of genetic techniques to combat mosquito vectors. One such technique involves introducing large numbers of sterile male mosquitoes in endemic areas. Vaccines against malaria are in active development. The fact that individuals repeatedly infected with the disease eventually become immune provides good reason to think that an effective vaccine is possible.

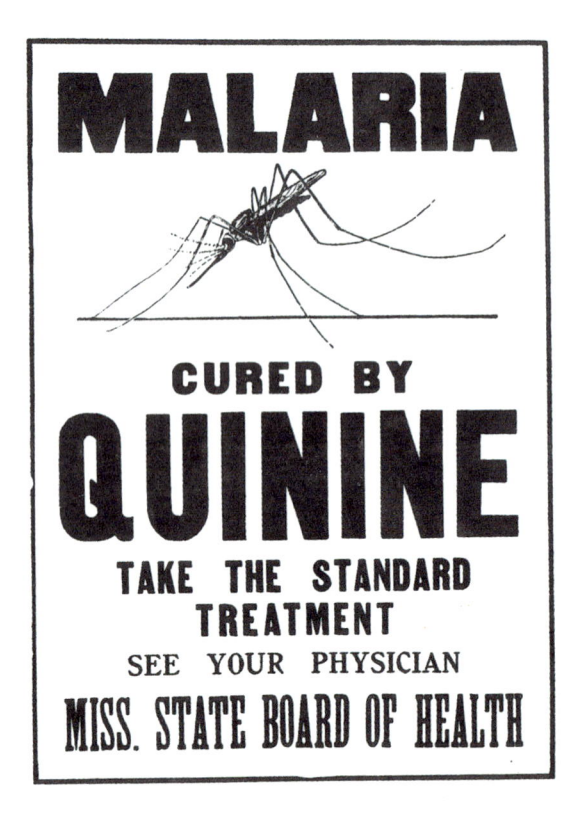

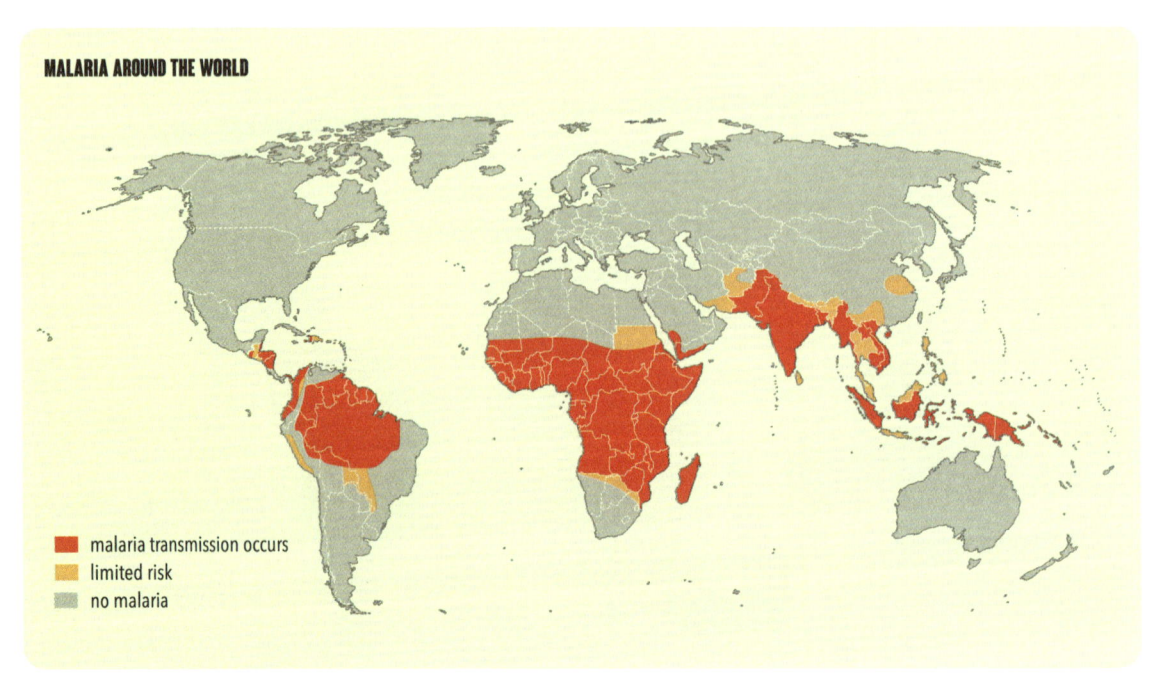

MALARIA AROUND THE WORLD

- malaria transmission occurs
- limited risk
- no malaria

Above top: Public health poster touting malaria treatment.

Opposite above: The use of nets to prevent mosquito bites.

Opposite below: Scientists in Ghana working to control malaria.

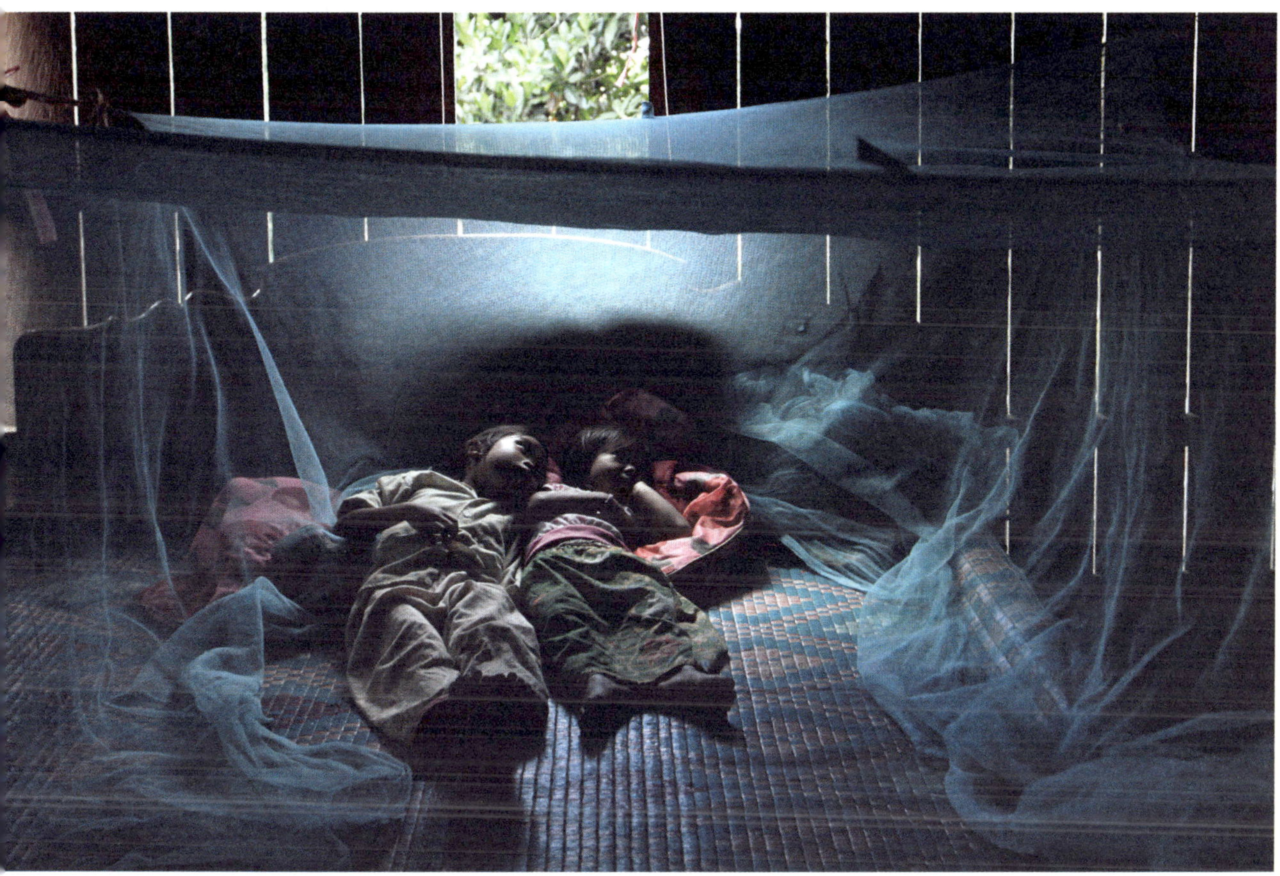

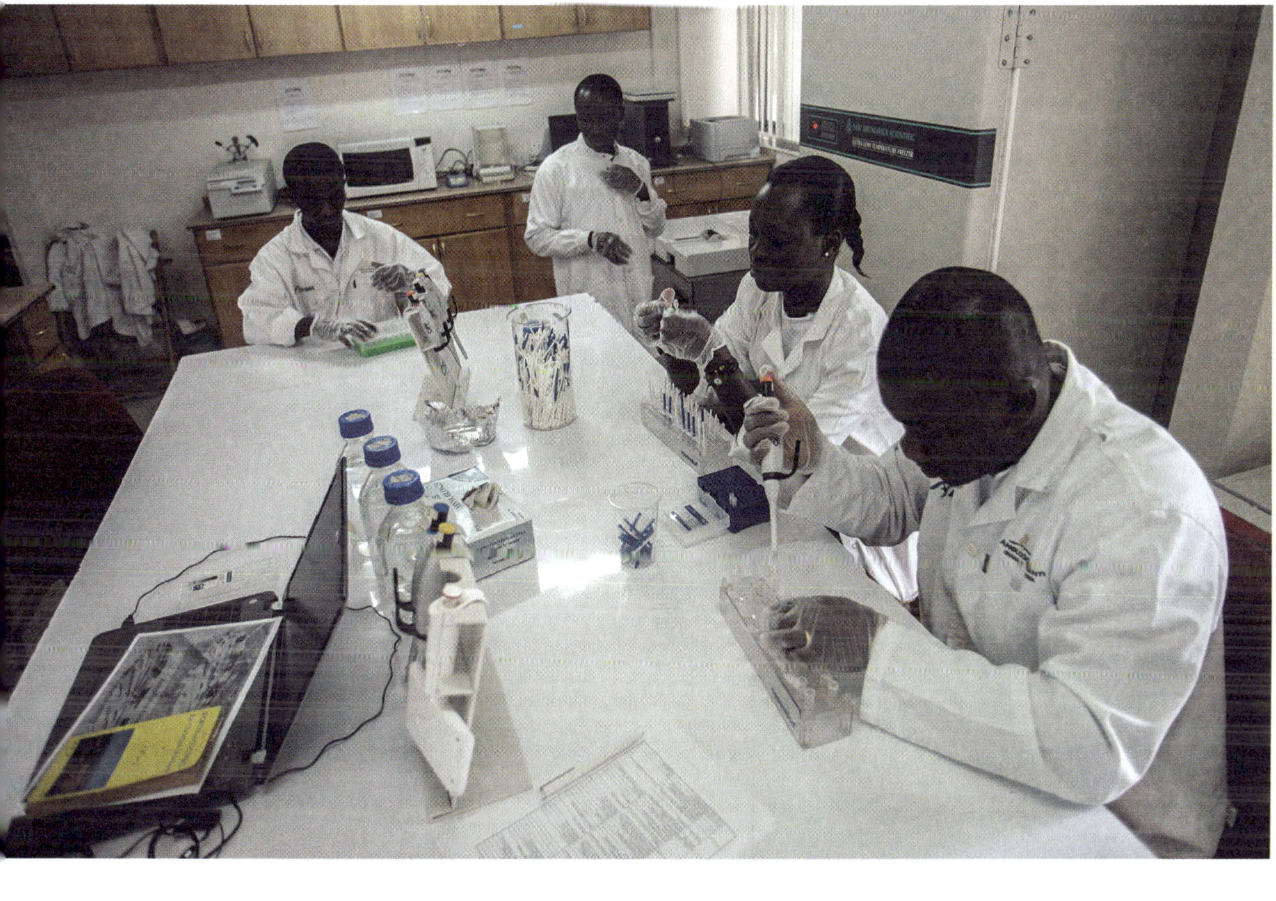

25 STIS: SEXUALLY TRANSMITTED INFECTIONS

DISEASES OF THE NIGHT

Sexually transmitted infections (STIs) are exactly what their name implies: infectious diseases transmitted by sexual activity. Many remain asymptomatic or cause only mild symptoms for a time, which contributes to a higher rate of transmission. Some can also be transmitted from mother to baby. More than two-dozen STIs have been recognized, with most due to bacteria, viruses, and parasites.

This chapter focuses on STIs other than HIV/AIDS and HPV/cervical cancer, which are covered separately. Excepting HIV/AIDS, it is estimated that approximately 1 billion people are infected with some form of STI, causing over one hundred thousand deaths per year. In the US, it is estimated that approximately 20 million new cases of STIs are diagnosed per year.

Broadly speaking, the most reliable way to prevent the transmission of sexually transmitted disease is to avoid sexual activity. Other strategies include decreasing the number of sexual partners, barrier contraception methods, especially condoms, and the use of vaccination to prevent hepatitis B and HPV infection. Many STIs are curable with antibiotics.

SYPHILIS

Syphilis is caused by the bacterium *Treponema pallidum*, which has a spiral shape and is often referred to as the spirochete. Initially, there is an ulceration at the site of infection, but later a more widespread rash may develop, followed in some years or even decades later by cardiac and neurologic disorders that can progress to death.

It is estimated that about 50 million people are infected with syphilis, with about 6 million new cases and 100,000 deaths per year. Fortunately, a single dose of a common antibiotic such as penicillin can cure the initial infection. After the disease progresses to late stages, a more prolonged course of treatment is usually necessary.

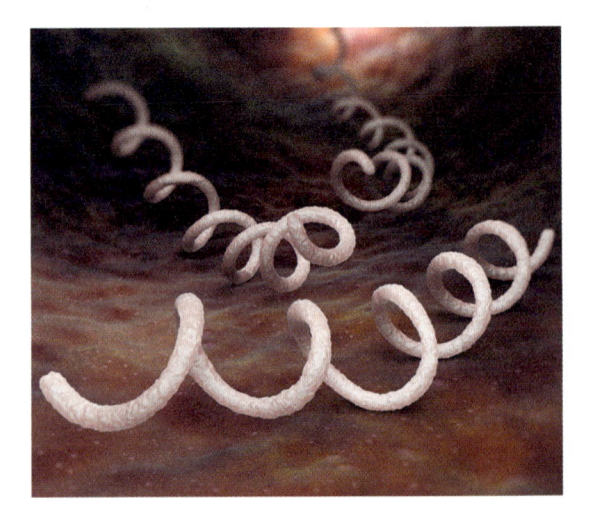

Above: *Treponema pallidum*, the bacterium that causes syphilis.
Opposite: American public health poster targeting syphilis.

STIs: Sexually Transmitted Infections

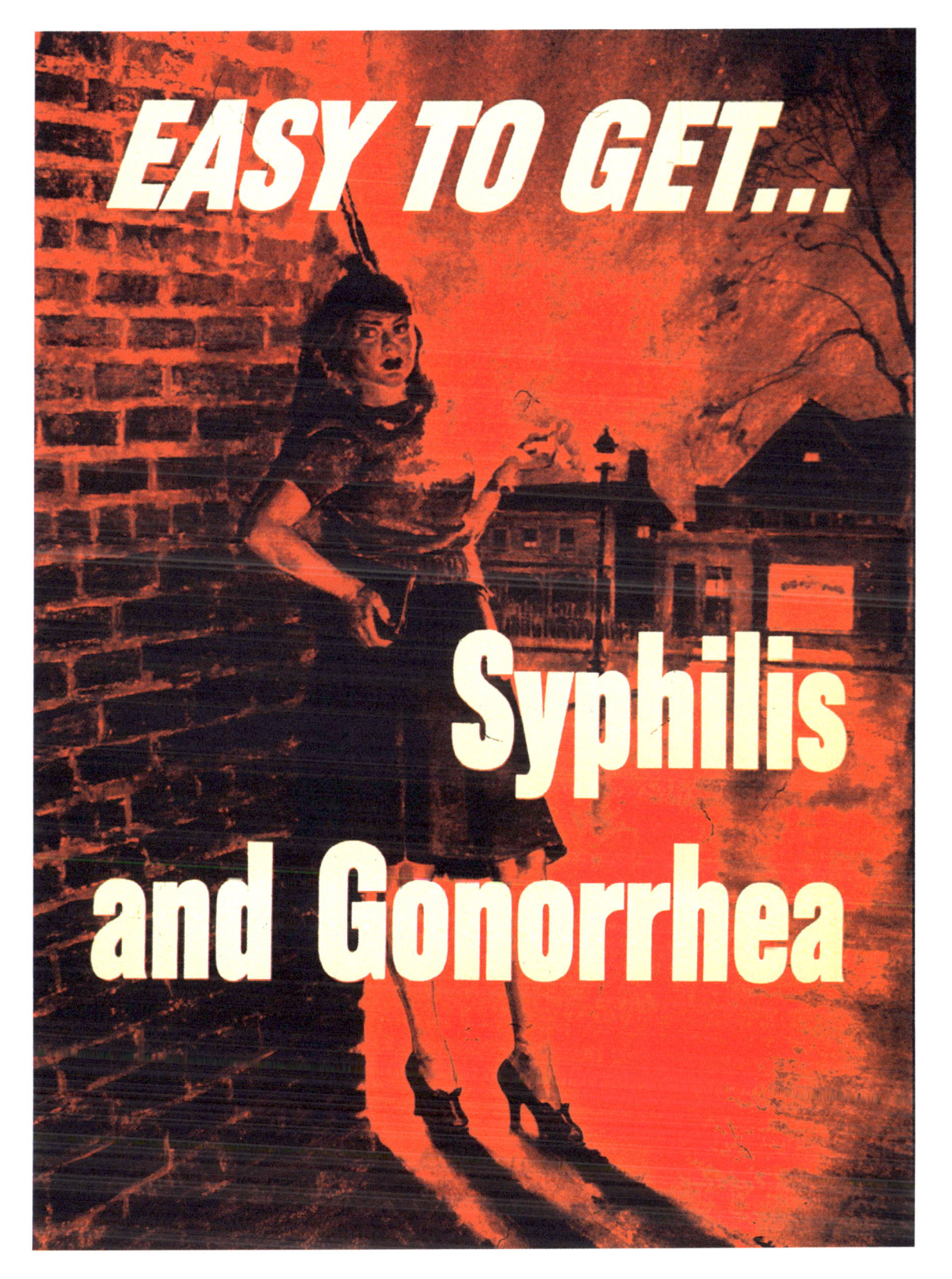

EASY TO GET...

Syphilis and Gonorrhea

The origin of syphilis is unknown, but it appears to have been present in the Americas prior to the arrival of Europeans and may have been carried back to Europe by early explorers. The first reported cases of the disease occurred in Italy in the late fifteenth century during a French invasion, leading some to call it the French disease.

Even before the spirochete was identified, it was generally understood that syphilis is an infectious disease that is transmitted sexually, and its wide prevalence led to further stigmatization of promiscuity and prostitution. Some of the compounds used to treat the disease, such as those containing mercury, may have inflicted as much harm as the disease itself.

The responsible bacterium was identified in 1905, and the first effective treatment, containing arsenic and known as Salvarsan, was produced soon thereafter. It represented one of the first effective chemical treatments of an infectious disease, but was replaced by penicillin when the latter became widely available in the 1940s.

One of the darkest chapters in the history of syphilis were the Tuskegee and Guatemala syphilis studies, undertaken by the US in the mid-twentieth century. In the former, poor black men with syphilis were left uninformed and untreated, and followed for decades to observe the progression of the disease. In Guatemala, subjects were infected with the disease without their consent.

GONORRHOEA

Gonorrhoea is caused by the *gonococcus* bacterium. In men, it causes painful urination and a penile discharge. In women, it causes painful urination, vaginal discharge, and pelvic inflammatory disease, an infection of the uterus and fallopian tubes. As many as 100 million new infections are thought to occur each year. Antibiotics are effective in curing the disease, but resistant infections are increasing.

The disease got its name from Greek roots meaning seed, as in gonad, and flow, reflecting the discharge it causes from the sexual organs. Some scholars suspect that descriptions of gonorrhoea are found in the Bible. It may have acquired one of its first names, "the clap", from the French, Le Clapier, an area of Paris where brothels were found and where the disease was often spread.

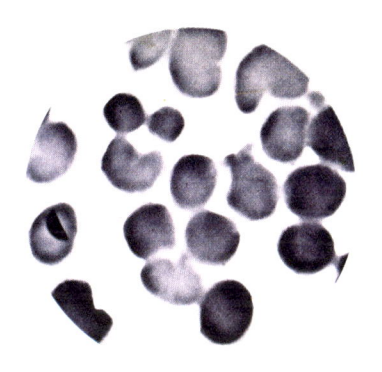

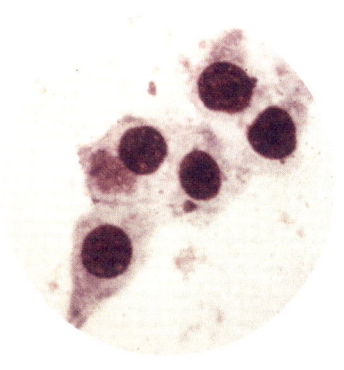

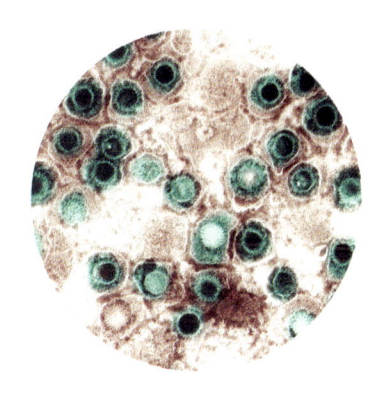

Laws apparently seeking to reduce the transmission of gonorrhoea can be found dating back to the twelfth century. The organism was first identified in 1879 by Albert Neisser (1855–1916), hence its scientific name, *Neisseria gonorrhoeae*. Early treatments included mercury compounds and silver nitrate. Penicillin was effective for decades, but in the 1980s penicillin-resistant bacteria began to appear.

CHLAMYDIA

Chlamydia, a term that refers to both an infectious organism and the STI it causes, is a bacterium, *Chlamydia trachomatis*. Many infected patients exhibit no symptoms, but women may experience burning with urination and vaginal discharge. Complications in women include pelvic inflammatory disease, which can result in infertility and increases the risk of ectopic pregnancy.

It is estimated that about 60 million new cases develop each year. Like gonorrhoea, chlamydia is not usually a lethal disease, but in developing parts of the world it was for many years a common cause of blindness, due to the spread of infection to the eyes. The disease can also be passed from mother to baby, causing problems such as pneumonia and eye infections.

Chlamydia gets its name from a Greek root meaning "cloak". The causative organism was discovered in 1907. Its one natural host is the human being, and it survives only within cells, making it an obligate intracellular bacterium. However, it is susceptible to commonly prescribed antibiotics, and early treatment can prevent the development of complications.

HERPES

Among the herpes viruses, *herpes simplex* virus can cause both oral herpes, blisters on the lips or mouth, as well as genital herpes, blisters on the sexual organs. These lesions often break open and form ulcers, but they generally heal within several weeks. Many patients experience recurrent attacks that decrease in frequency and intensity over time.

Two types of herpes viruses are responsible. *Herpes simplex* virus type 1 (HSV-1) usually causes oral lesions, while HSV-2 generally causes genital lesions. Once a patient is infected, the virus persists in sensory nerves throughout life, often reactivating during periods of stress. Antiviral drugs may help, but the disease is not curable, and no vaccine is available.

It is estimated that at least two-thirds of adults worldwide are infected, many of whom are unaware. The disease can be particularly problematic in patients with impaired immune systems, such as those with HIV/AIDS, who may develop inflammation of the brain and its covering membranes (encephalitis and meningitis), as well as inflammation of the liver (hepatitis).

Herpes comes from the Greek meaning "creeping", which is thought to reflect the spreading of the blisters. It has been known for thousands of years, and the genital form is thought to have jumped from chimpanzees to humans more than 1 million years ago. The viral cause was identified in the 1940s, and multiple antiviral medications have been developed since to control the disease.

Opposite: Students working in the laboratory at the Tuskegee Institute in Alabama.
Above, from top: The bacteria that cause gonorrhea; *Chlamydia trachomatis*; electron micrograph of Herpes virus.

26 PENICILLIN

Bacterial infections of one kind or another had devilled humanity throughout its history, and perhaps the single most dramatic step in the conquest of disease-causing bacteria occurred with the discovery and production of penicillin. A Scottish physician, Alexander Fleming (1881–1955), made the discovery quite by accident in 1928.

The son of a farmer and his second wife, Fleming graduated with honours from medical school in 1906. He then undertook bacteriological research before serving in the First World War, after which he became a professor of bacteriology in London. Having seen many soldiers die of wound infections, Fleming began investigating antibacterial substances.

FLEMING'S UNEXPECTED DISCOVERY

His first discovery was lysozyme, an enzyme with antibacterial properties found in mucus, tears, and saliva. In the midst of his investigations of *Staphylococci*, a common form of disease-causing bacteria, he went away on holiday. When he returned, he found that his Petri dishes had become colonized with a fungus. He also noticed that there were no staphylococcal colonies in the vicinity of the fungi.

Fleming then began growing the fungus, the common mould *Penicillium* (from the Latin *penicillus*, for paintbrush, for its brush-like fruiting bodies), and isolating the antibacterial substance it produced. Instead of calling it "mould juice", he named it penicillin. He then showed that penicillin could inhibit the growth not only of *Staphylococci* but also of many other types of bacteria.

However, Fleming found *Penicillium* difficult to grow and penicillin even harder to isolate, and he also came to believe that the substance was too unstable to last long enough in the human body to serve as an effective medication. To cap it all off, he also determined that it would be impossible to produce it in sufficient quantities to be useful in patient care.

Below: Growth of the mould that produces penicillin, over the course of 10 days.

Opposite: Bacterial colonies at a distance from penicillin-producing mould.

Penicillin

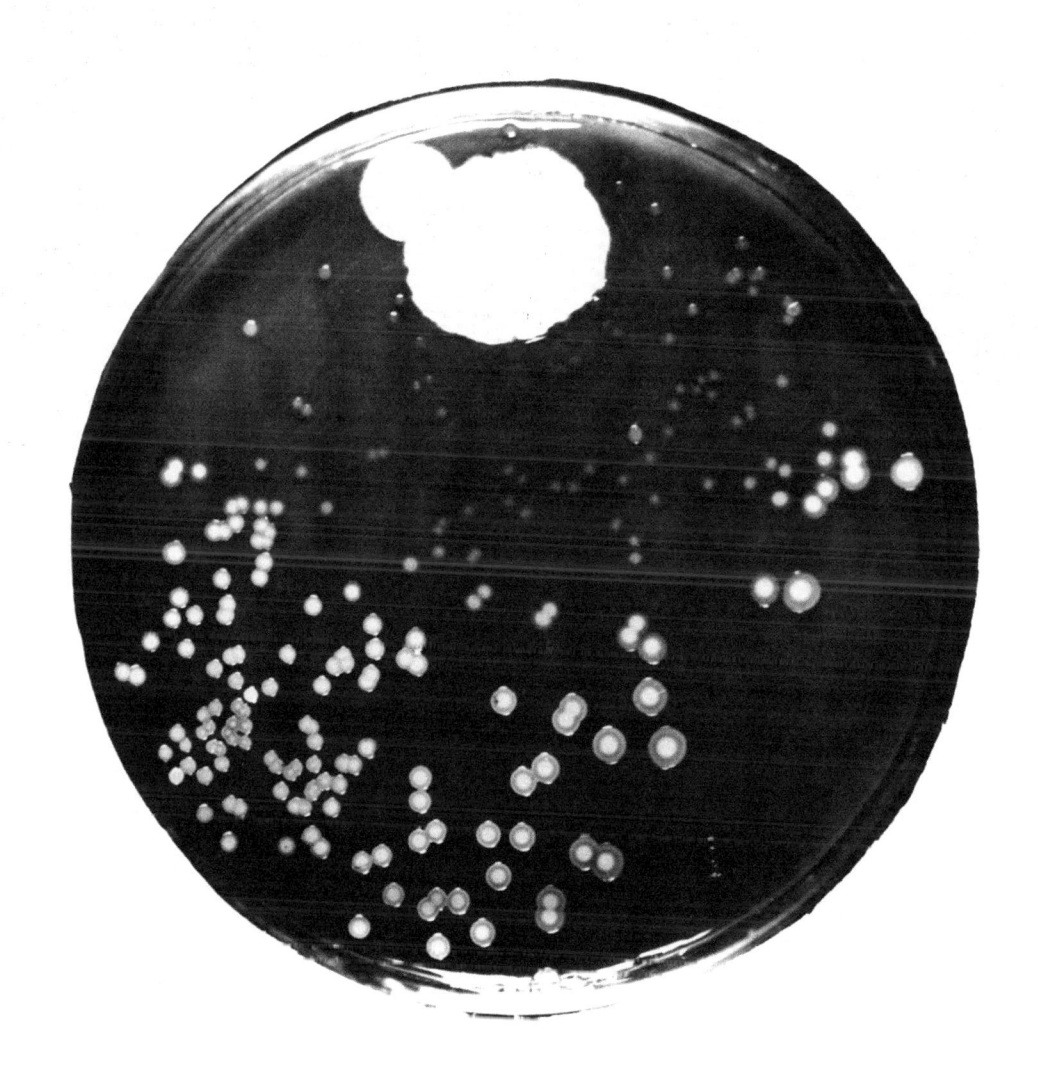

The beginning of Penicillin
Alexander Fleming

THE WORK CONTINUED

Fortunately, other researchers became intrigued by Fleming's discovery. Howard Florey (1898–1968) was an Australian pathologist who attended Oxford University on a Rhodes Scholarship. Ernst Chain (1906–79) was a German biochemist who fled the Nazis in 1933 and joined Florey at Oxford in 1939. Other key figures in the penicillin story were biochemists Norman Heatley and Edward Abraham.

Intrigued by the technical challenges Fleming had identified, Florey and Chain set to work on Fleming's discovery, investigating its effects in living organisms. In 1940, they showed that it could cure bacterial infection in mice. In 1941, they administered it to a policeman with a severe skin infection on his face. The infection responded, but they ran out of penicillin, and he died.

Florey and Chain began work on large-scale production of penicillin. With the Second World War underway, British resources were largely consumed by the conflict, so Florey travelled to the US to see if he could interest pharmaceutical companies in a drug that might cure soldiers' bacterial infections. By 1942, the US company Merck was mass-producing the drug, and by 1944, doses numbered in the millions.

One challenge that soon emerged was the high rate of the drug's clearance from the body. Within just hours of administration, four-fifths would be excreted by the kidneys. In the early days of penicillin, it was so highly prized that patients' urine would be collected so that it could be extracted and reused.

A short-lived solution to this problem emerged when a drug used to treat gout, probenecid, turned out to competitively inhibit penicillin excretion. By giving the two drugs together, penicillin would remain in the body longer, prolonging its effect. Eventually, however, the ability to produce huge quantities of penicillin and the development of other antibiotics reduced probenecid's use.

FURTHER ADVANCES

In 1942, Edward Abraham proposed a chemical structure of penicillin, and this was confirmed by Dorothy Hodgkin (1910–94). Born in Egypt to British parents serving in the educational ministry, Hodgkin showed a passion for chemistry, and on her sixteenth birthday her mother gave her a book on x-ray crystallography. She studied at Oxford and Cambridge, receiving her PhD in 1937.

Returning to Oxford, she conducted x-ray crystallography studies of many compounds before setting to work on penicillin. Despite developing rheumatoid arthritis that caused severe pain and deformity in her hands and eventually landed her in a wheelchair, she was remarkably productive throughout her life. She determined the structures of penicillin, vitamin B_{12}, and insulin.

The key chemical component of penicillin is a structure called the beta-lactam ring, which damages the cell wall of bacteria. As bacteria grow and divide, they disassemble and reassemble their cell walls. The beta-lactam ring of penicillin binds to an enzyme that works to bind the components of the cell wall together, quickly killing the cell.

Unfortunately, some bacteria produce an enzyme called beta-lactamase, which breaks the beta-lactam ring of penicillin, rendering it and related antibiotics ineffective. Chain and Abraham found the enzyme even before penicillin had entered commercial production, which indicates that it evolved in bacteria as a defence against penicillin. Widespread antibiotic use has increased its prevalence.

LEGACY AND CONTINUING RESEARCH

Research on penicillin gave rise to many progenies. In 1961, ampicillin was introduced, offering a broader antibacterial spectrum. Following it was methicillin, a synthetic penicillin that resisted beta-lactamase. Derived from a different fungus were the cephalosporins, introduced first in 1964 in multiple generations. They are effective against an even wider spectrum of bacteria.

Many of the scientists who worked on penicillin were rightly rewarded. In 1945, Fleming, Florey, and Chain shared the Nobel Prize in Physiology or Medicine for their work, and In 1964, Dorothy Hodgkin received the Nobel Prize in Chemistry, partly for determining penicillin's molecular structure. Florey considered patenting the drug, but determined it would be unethical to do so.

Penicillin spared many Second World War soldiers who would have died or suffered amputations from infected wounds. It was also widely used by the military to treat sexually transmitted diseases such as gonorrhoea. After the war, it became available to civilians. Some have suggested that, since its discovery, penicillin may have saved as many as 200 million lives.

> **SOME HAVE SUGGESTED THAT, SINCE ITS DISCOVERY, PENICILLIN MAY HAVE SAVED AS MANY AS 200 MILLION LIVES**

Opposite: Alexander Fleming.
Right: Biochemist Ernst Chain.
Above: British chemist Dorothy Hodgkin.

27 ATTEMPTS TO ERADICATE INFECTIOUS DISEASE

In most cases, battles against infectious diseases have been just that: battles. For a long time, human beings tended to fare rather poorly, and many lives were lost as diseases such as cholera, bubonic plague, and syphilis took their toll. In other cases, such as John Snow's battle with the Broad Street pump or Alexander Fleming's discovery of penicillin, human beings seemed to emerge victorious.

Yet the outcomes of such battles were always mere shifts in a balance from which both sides, pathogen and host, lived to fight another day. The idea of winning the war – achieving a complete and final victory – appeared unimaginable. For any microbe to win, humanity would need to be exterminated, and for humanity to win, it would need to drive a pathogen to extinction.

Human beings have attempted such a victory before. For example, hookworm, yellow fever, yaws, and malaria have appeared within the crosshairs of such campaigns, yet while great progress has been made against them, the pathogens persisted. Malaria, for instance, could rear its ugly head again if vigilance against its mosquito vector were ever to slacken.

SMALLPOX

The first and only successful effort by humans to eradicate a human infectious disease was the campaign against smallpox. The vaccination programme began in the 1960s, and within just half a decade or so, the number of countries with endemic smallpox dropped from about thirty to only five. But as the number dropped, the degree of difficulty increased.

For example, two African nations presented special challenges. Ethiopia was embroiled in civil war, hampering access of public health officials to different portions of the country. And Somalia had a poor disease-monitoring system that tended not to report cases until they reached epidemic level. But finally, in 1977, the last naturally occurring case of smallpox was identified.

POLIO

Polio, by contrast, serves as a case study in the failure of efforts to eradicate an infectious disease. Humanity may still eliminate polio, but predictions of impending eradication are now delayed by decades. To understand the difficulties besetting such eradication efforts, it is necessary to know something about the disease and the virus that causes it.

The polio virus is an RNA virus, and was first isolated in 1904 by Karl Landsteiner (1868–1943), who had distinguished the main human blood types in 1900, for which he received the Nobel Prize in Physiology or Medicine in 1930. Rosalind Franklin (1920–1958), an important contributor to the discovery of DNA's double-helix structure, used x-ray diffraction to show the virus's structure.

The virus is transmitted by a faecal–oral route, often through contaminated food or water. Once infected, people may transmit the microbe even if they are

Opposite: A child with post-polio syndrome learns to walk with crutches.

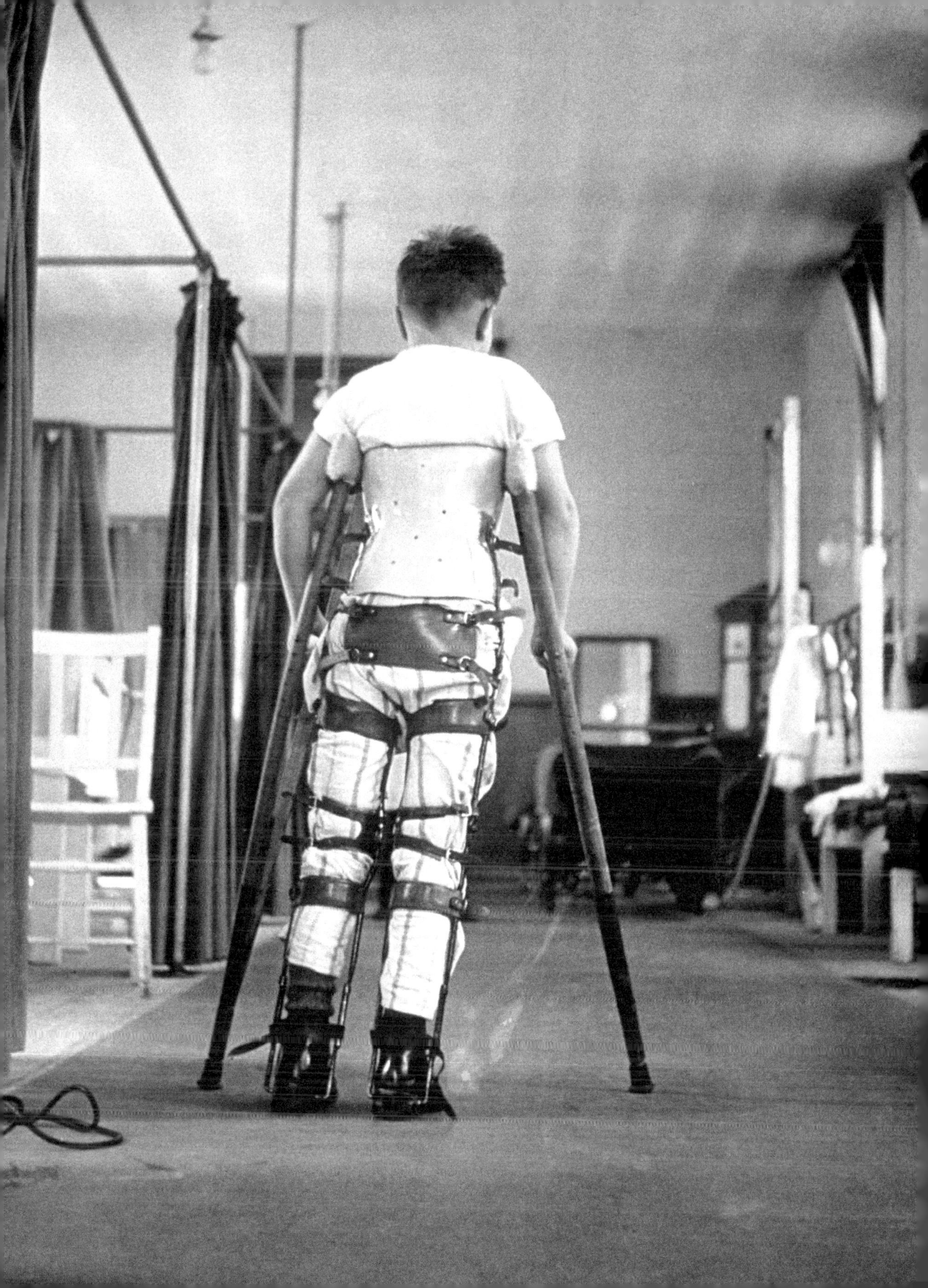

asymptomatic for as long as six weeks. Nearly three-quarters of people manifest no symptoms, but about a quarter develop fever and sore throat, and a small percentage muscle weakness and/or paralysis.

Many patients fully recover, but years later a few develop post-polio syndrome, with progressive muscle weakness and paralysis. In some cases, the paralysis may affect the ability to breathe, which explains why, in the twentieth century, many patients were placed in "iron lungs", which could create periodic negative pressure and expand the chest, drawing air into the lungs.

In the 1 per cent of patients who develop paralysis, the virus spreads along nerve pathways into the spinal cord, where it can destroy the motor neurons that stimulate muscles to contract. The older the patient, the more likely is the development of paralytic polio. Children often suffer paralysis in the lower extremities, while in adults it tends to involve more of the body, including muscles of respiration.

The peak incidence of paralytic polio occurred in Europe and the United States in the 1950s, with the worst US epidemic affecting 58,000 people in 1952. About 20,000 people were left with paralysis. In response, grass-roots philanthropic efforts were launched, including the March of Dimes, which attracted attention in part because US President Franklin Roosevelt suffered post-polio syndrome.

Vaccines against polio were developed in the 1950s. Jonas Salk (1914–95) developed an inactivated virus version that was administered by injection, producing antibodies against all types of the polio virus in more than 99 per cent of those immunized. Albert Sabin (1906–93) developed a live, oral polio vaccine that remains confined to the intestines and produces antibodies in 95 per cent of recipients.

During the 1960s, Sabin's vaccine became the preferred version and was deployed worldwide. However, the Sabin vaccine is not perfect, and in rare cases, about one in a million, the attenuated virus can revert to a paralytic form.

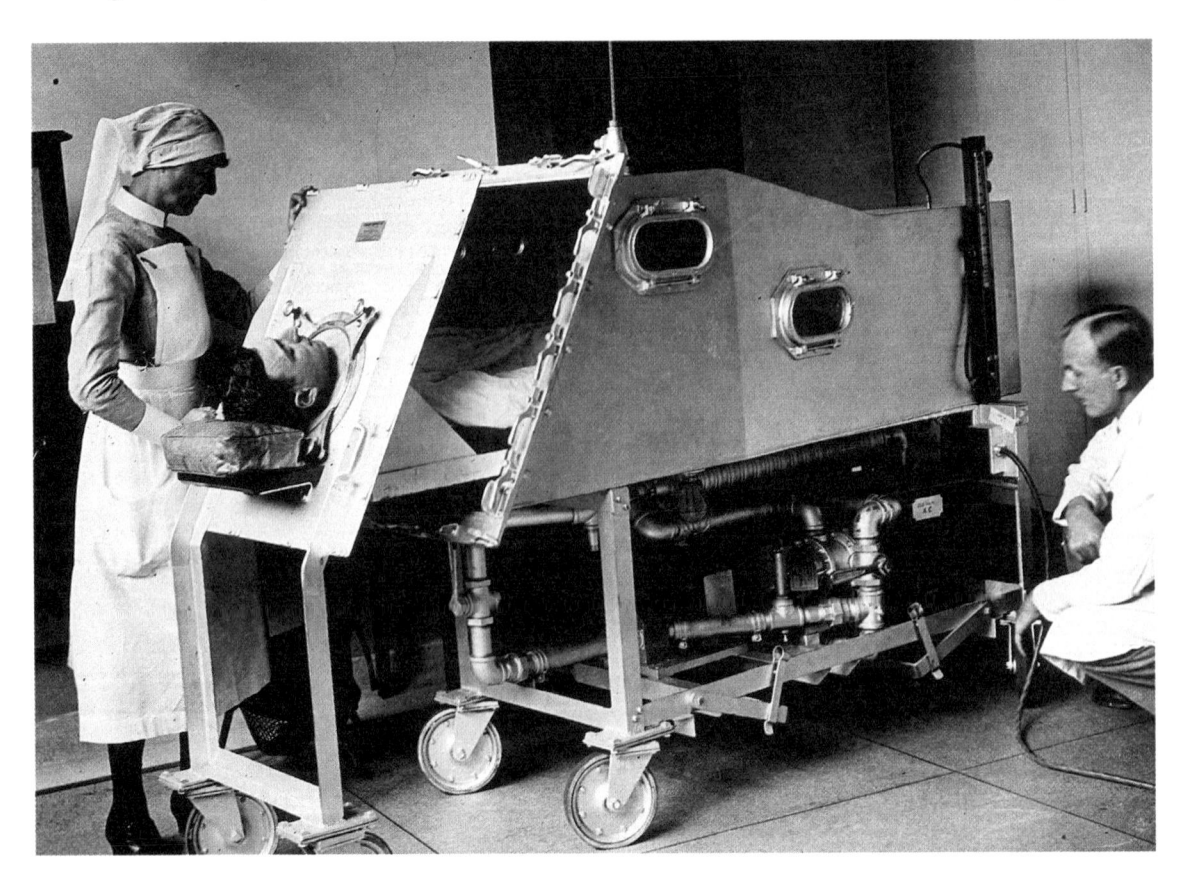

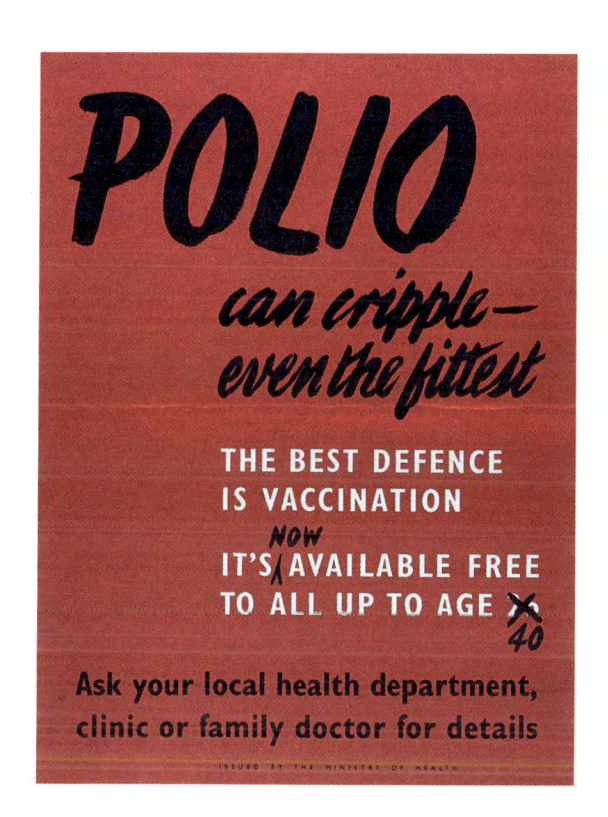

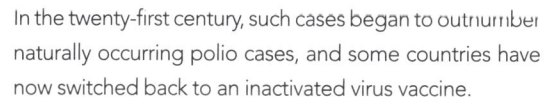

Opposite: A patient in an early mechanical respirator.
Above and right: A public health poster encouraging polio immunization; the polio vaccine; Albert Sabin.

In the twenty-first century, such cases began to outnumber naturally occurring polio cases, and some countries have now switched back to an inactivated virus vaccine.

Efforts to eradicate polio were initially very promising. For example, the last case of naturally occurring polio in the Americas occurred in the early 1990s, and Europe became polio free in the early 2000s. However, polio cases are still occurring in Afghanistan, Pakistan, and Nigeria, where approximately a hundred cases per year have been documented in each country.

Such countries present difficulties. These include rugged terrain, political unrest, and armed conflicts that make it difficult to reach all portions of a country. Some groups regard the activity of immunization teams as a threat to their authority or a covert effort to spy on them, and rumours have even circulated that vaccination represents an insidious plot to sterilize local populations.

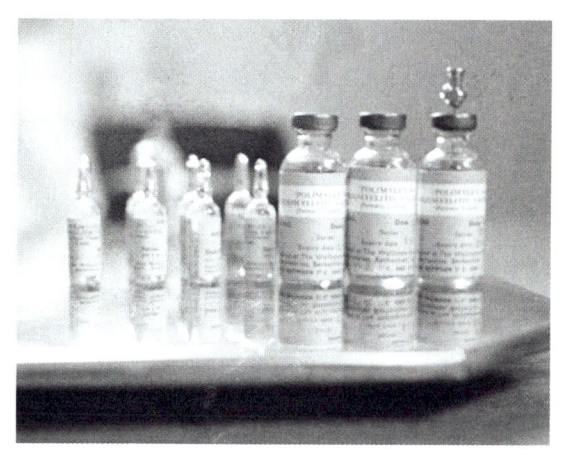

BARRIERS TO ERADICATION

Often the chief barriers to eradication of an infectious disease for which an effective vaccine exists, such as polio, are unpredictable. These include natural disasters such as earthquakes, floods, and famines, as well as political and military conflicts that can produce hundreds of thousands or even millions of refugees. Governments and aid organization bureaucracies can prove problematic, as well.

Another potential difficulty with disease eradication is its potential to destabilize existing health infrastructure. When a large international immunization effort enters a country or region, it may co-opt existing health professionals and institutions for its own purposes, leaving fewer resources to attend to routine healthcare needs, such as delivering babies and setting broken bones.

28 HIV/AIDS

Before 1981, no one had ever heard of Acquired Immune Deficiency Syndrome (AIDS), yet within just 20 years it ranked as the world's seventh most common cause of death, taking the lives of about 1.5 million people per year. For this reason, AIDS is considered a pandemic, an actively spreading infectious disease that is found throughout the world.

AIDS was first identified in the United States in 1981. The Centers for Disease Control (CDC) *Morbidity and Mortality Weekly Report* contained a report of five cases of rare *Pneumocystis* pneumonia among homosexual men in Los Angeles. Soon, similar clusters were found in other cities, along with cases of a rare cancer, Kaposi's sarcoma, in comparable populations.

By the next year, a new syndrome was postulated, Gay-Related Immune Deficiency or GRID. It soon became apparent, however, that cases were not confined to homosexual men. Similar infections were seen among intravenous drug users and haemophiliacs. By the end of 1982, a new name for the disease had been coined: AIDS.

Just a year later, in 1983, a team of investigators at the Pasteur Institute in Paris announced that they had isolated a new retrovirus from AIDS patients. An investigator at the National Cancer Institute announced that he had independently isolated the virus, but by 1985 it had become apparent that the American virus was from the same patient the French investigators had reported.

By 1986, the virus was renamed HIV, for Human Immunodeficiency Virus. The 2008 Nobel Prize in Physiology or Medicine was awarded to Luc Montagnier and Françoise Barré-Sinoussi for the discovery of HIV. Once the causative agent was identified, it was possible to study the virus and begin to develop antiviral medications.

ORIGINS OF THE VIRUS

Where the virus came from is a matter of considerable investigation and debate. It seems likely that HIV emerged from a similar virus that infected non-human primates such as chimpanzees and gorillas in west central African forests. It appears to have moved from apes to humans sometime in the late nineteenth or early twentieth century, with subsequent urbanization accelerating its transmission.

Many scientists suspect that the jump to humans resulted from regional bushmeat practices – specifically, when a hunter or butcher was bitten or otherwise exposed to an animal's blood. This theory is supported by the fact that a small percentage of people in this African region harbour antibodies to the simian immunodeficiency virus.

Above: Electron micrograph of HIV virus.
Opposite: American public health poster warning against AIDS.

A man who shoots up can be very giving.

He can give you and your baby AIDS.

THE WASHINGTON AREA COUNCIL ON
ALCOHOLISM AND DRUG ABUSE, INC.
1232 M Street, N.W.
Washington, D.C. 20005

Most babies with AIDS are born to mothers who shot drugs or who sleep with men who have.

Babies with AIDS are born to die.

If you're thinking of having a baby you and your partner need to get tested for AIDS. Only get pregnant when you're sure both of you aren't infected. Until then help protect yourself and your partner by using condoms.

And if your man shoots drugs, help him get into treatment now. It could save three lives, his, yours and your baby's.

STOP SHOOTING UP AIDS.
GET INTO DRUG TREATMENT.
CALL 1-800 662 HELP.

A Public Service of the National Institute on Drug Abuse, Department of Health and Human Services

IMPACT

Today, it is estimated that nearly 40 million people around the world are infected with HIV, resulting in nearly 1 million deaths per year. Approximately half of these individuals live in central and eastern Africa. As many as 32 million people worldwide are thought to have died of AIDS, and it is estimated that between 1.5 and 2 million people are being infected each year.

Individuals infected with HIV generally experience a short-lived flu-like illness, but some experience no symptoms at all. Then, for a prolonged period of time, the disease has no symptoms. Years later, patients begin developing weight loss and increased numbers of infections, including types generally seen only in patients with damaged immune systems, such as *Pneumocystis* pneumonia.

HIV is generally spread by sexual contact, blood transfusions containing the virus, shared hypodermic needles, and during pregnancy and birth. Hence, means of preventing transmission include safe sex, particularly the use of condoms, avoiding needle sharing, and medications that can markedly reduce or eliminate the presence of the virus in the blood of infected patients.

THE CONTINUING STRUGGLE

Despite the fact that HIV is one of the most studied viruses in history, it remains a very hard nut to crack. First, it is a retrovirus, meaning that its genetic material is RNA, not the DNA found in most organisms (roughly speaking, DNA makes RNA makes proteins). One of the enzymes for which its genes encode is reverse transcriptase, which makes a DNA copy of its RNA that can be inserted into the chromosomes of infected cells.

Once the virus-produced DNA is incorporated into host cells, it can remain there for years and even decades. At some point in the future, it becomes activated, directing the host cell to make more copies of the virus, including the RNA necessary for their chromosomes. Because the viral DNA remains safely inside of host cell chromosomes, it is largely impossible to cure HIV infection.

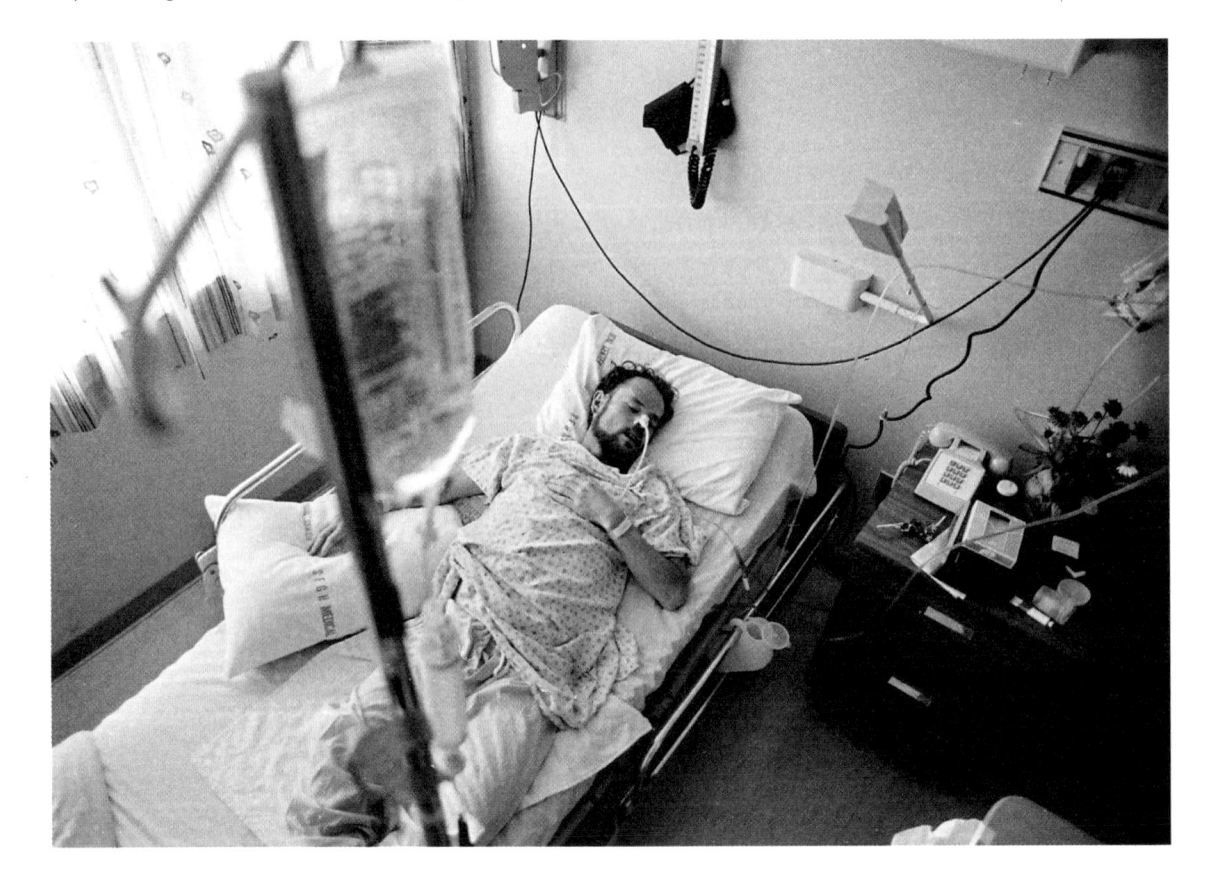

> **TODAY, IT IS ESTIMATED THAT NEARLY 40 MILLION PEOPLE AROUND THE WORLD ARE INFECTED WITH HIV, RESULTING IN NEARLY 1 MILLION DEATHS PER YEAR.**

Another difficulty in combating HIV is the fact that it has a very short life cycle – just a day or two between initial infection and the production of viruses that can infect other cells. In addition, it is sloppy – that is, its conversion of RNA to DNA is not very accurate, meaning that errors or mutations occur often.

As expected, most mutations in HIV are harmful to the virus, producing "offspring" viruses that are less suited to survival than the "parent". But occasionally, a mutated virus turns out to be better adapted than its parent, conferring a survival and reproduction advantage. Such a mutation might enable HIV to better avoid host immune response or to resist an antiviral drug.

The difficulty of curing HIV is compounded by the types of cells it infects, so-called helper T cells. These cells play important roles in helping to coordinate the immune system's response to bacterial and fungal infections. Initially, helper T cell numbers in the blood remain normal, but over years their levels drop, rendering patients much more prone to infections and even some cancers.

Opposite: Photo of one of the early American patients hospitalized with HIV/AIDS.
Below: Protesters marching against the US Food and Drug Administration.

THERAPY NOT CURE

Broadly speaking, there are about a dozen steps in the life cycle of HIV, each of which could serve as a target for therapy. Single medications tend not to be effective for long, because they create an environment favourable to mutated versions of the virus that are drug resistant. As was the case for tuberculosis therapy, a breakthrough occurred when investigators utilized multiple drugs.

Highly active antiretroviral therapy (HAART) utilizes a cocktail of drugs that target two steps in the viral life cycle at different stages. These block the action of the reverse transcriptase enzyme, preventing the production of a DNA copy of the viral genome, and they also block viral proteases, enzymes that synthesize viral proteins. The probability of a single virus developing three or four mutations simultaneously is incredibly low.

Today, these multiple antiviral drugs can be combined into a single-pill form, making it much easier for patients to take their medications reliably. The benefits of HAART are numerous and include reducing viral levels in the blood to the point that they are undetectable, blocking progression of the infection to full-blown AIDS, and preventing virus transmission to non-infected individuals.

To physicians who cared for the first HIV-infected patients in the 1980s, this represents a radical transformation in HIV/AIDS. A disease that was once thought to be uniformly fatal, although it admittedly remains still essentially incurable, has become a manageable chronic disease that patients can live with in a nearly normal state of health for decades after they become infected.

Right: A combination of medications used to treat HIV/AIDS.
Far right: A researcher in the Pasteur Institute, where HIV was first identified.

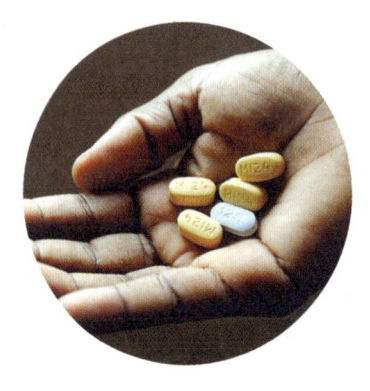

HIV/AIDS

29 IS PEPTIC ULCER DISEASE INFECTIOUS?

Almost a century after German scientist Max Pettenkofer downed a draught swirling with cholera in an attempt to prove that the organism did not cause the disease, Australian gastroenterologist Barry Marshall (b. 1951) was engaged in a similar demonstration with the opposite intended effect. Armed with Koch's postulates and a draught of another microbe, Marshall was attempting to prove that peptic ulcer disease, which afflicts tens of millions of people around the world, is infectious, by infecting himself.

The efforts of Marshall and his collaborator, Australian pathologist Robin Warren (b. 1937), would produce a paradigm shift in medicine's understanding of the cause of gastritis, an inflammatory condition of the lining of the stomach, and ulcers, which represent erosions of the wall of the stomach and first part of the small intestine. Marshall and Warren were not the first to observe microbes in the stomachs of ulcer patients, but they were the first to establish a causal relationship.

CHALLENGING ESTABLISHED THEORIES

To understand the magnitude of their achievement, it is important to understand the long-accepted view of the cause of peptic ulcer disease. Physicians around the world had long assumed that excess acid was the culprit. The human stomach naturally produces large quantities of hydrochloric acid (HCl), which both protects the rest of the digestive system from ingested microbes and assists in the normal digestion of proteins in foods such as meats.

Thanks to the secretion of HCl by some of the cells that line the stomach, the pH or hydrogen ion concentration in the stomach is about ten thousand times higher than that in the blood. It is so high, in fact, that many physicians categorically denied that any bacteria could survive in the stomach. Instead, they argued, peptic ulcers resulted from one of two causes: excessive gastric acid secretion or insufficient protection of the gastric lining by mucus, both of which could be exacerbated by stress.

For many years, the battle against peptic ulcer disease was guided by this paradigm. For example, ulcer patients were urged to consume bland, non-acidic diets, to take antacids that would buffer gastric acid, and more recently, to take medications that directly blocked gastric acid secretion. For decades in the twentieth century, patients with refractory disease even underwent surgical operations to remove the part of the stomach where most acid secretion occurs.

Below: Illustration of an ulcer in the stomach.
Opposite: Electron micrograph of *Helicobacter pylori*.

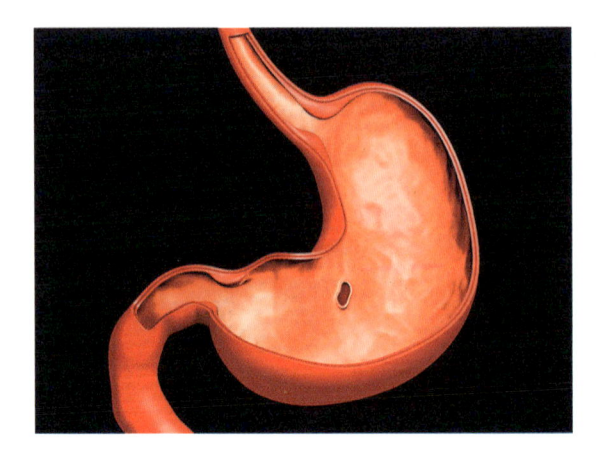

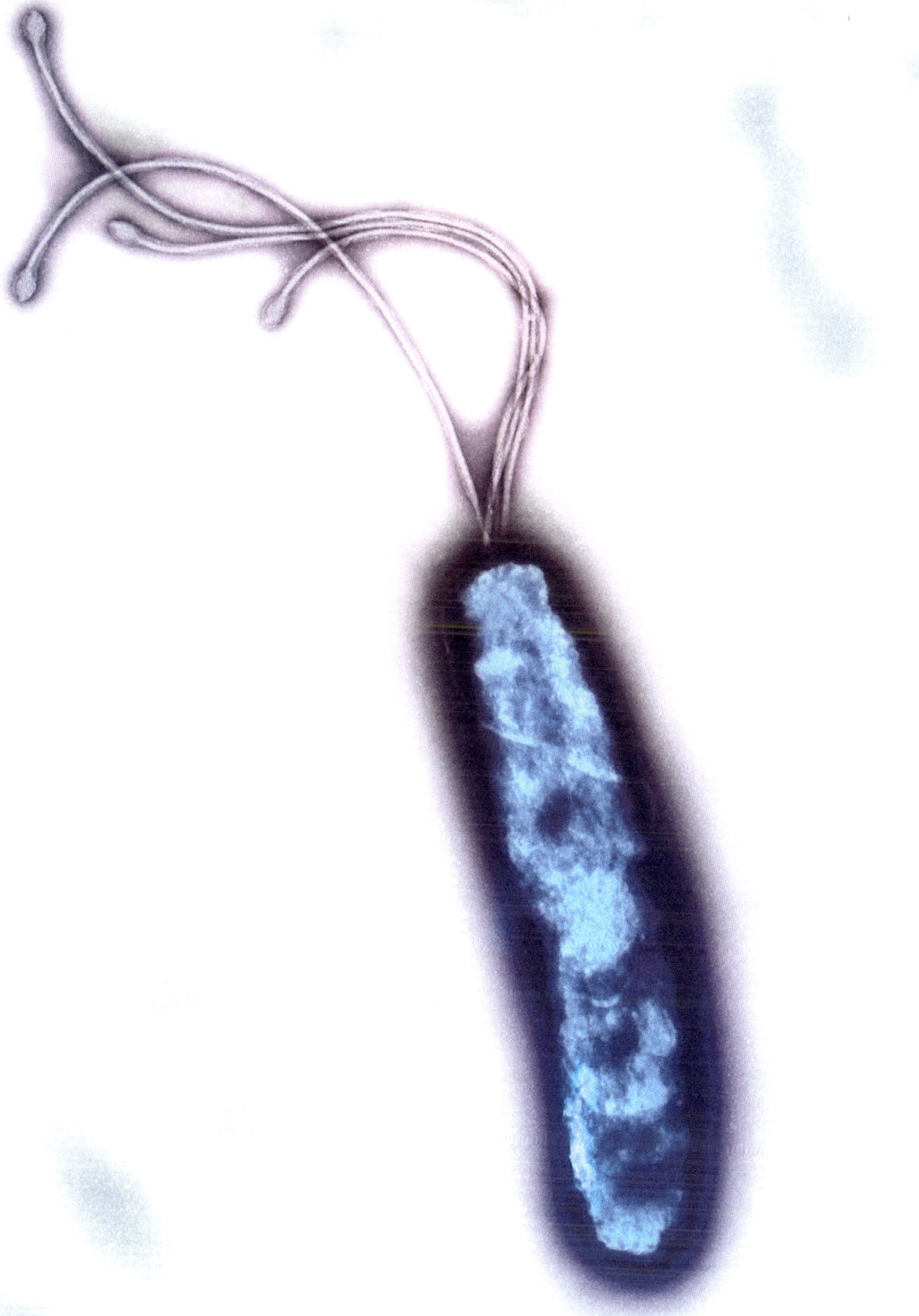

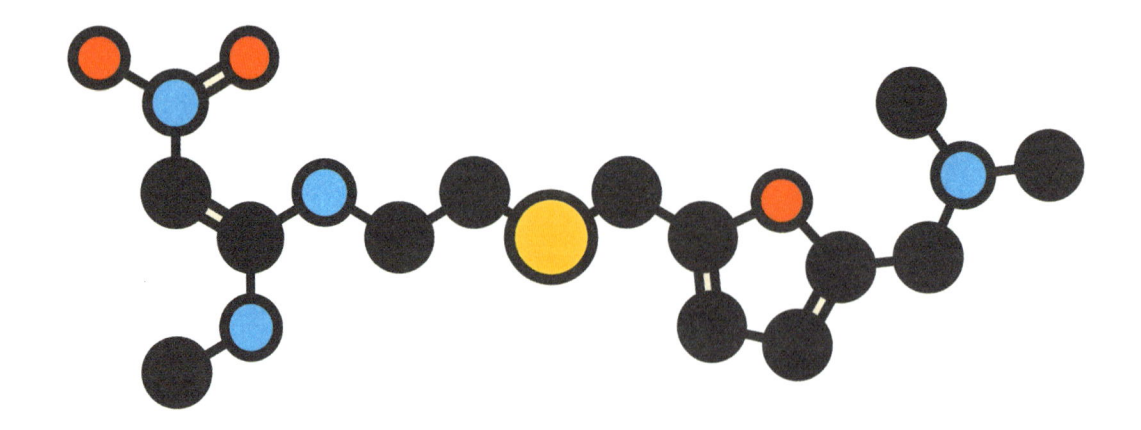

COLLABORATION

Marshall and Warren met in 1978 at Royal Perth Hospital, where Warren was a practising pathologist and Marshall was completing his fellowship. Marshall proposed that they work on a research project together, following up on Warren's observation several years earlier that biopsy specimens from gastritis patients revealed the presence of many curved bacteria. The number, location, and arrangement of the bacteria led Warren to believe that they must be associated with the disease.

When Warren shared his suspicions with gastroenterologists, however, their response was generally the same: "If these bacteria cause gastritis, why hasn't this been described before?" Ironically, the fact that the discovery had not already been made was preventing many physicians around the world from making it. Attempting to explain his own conviction in the face of such opposition, Warren later wrote, "I preferred to believe my eyes, not the medical textbooks or the medical fraternity."

In 1982, Marshall and Wright began studying around a hundred patients, attempting, as Koch would suggest, to culture the organism from their biopsy specimens. This turned out to be quite difficult to accomplish, in part because the first batch of samples grew out no bacteria. However, it turned out that laboratory technicians were disposing of the cultures after two days, which was standard practice for other cultures. When a culture was accidentally retained for four days, it grew out the bacteria.

MEDICAL RESISTANCE

As it turned out, the bacterium in question required more time than others to grow in culture. Later that year, Marshall presented his preliminary results at a local meeting, but he was met with scepticism. His colleagues were reluctant to abandon their long-held beliefs concerning the cause of peptic ulcer disease. Marshall and Warren later said that they might have given up but for their mutual support and encouragement from their wives.

Marshall later expressed suspicion that conflicts of interest may have delayed recognition of the work he and Warren were promoting. For one thing, pharmaceutical companies were generating billions of dollars in revenue producing drugs to decrease gastric acid. Moreover, gastroenterologists were generating considerable income by performing numerous endoscopies each week, a source of revenue that would decrease substantially if the disease could be cured once and for all.

In 1983, Marshall and Warren managed to get two letters published in *The Lancet*, which included both light microscope and electron microscope images of the organisms. Later in the year, Marshall presented their findings at an international meeting of microbiologists. This audience was less difficult to convince than the physicians. Finally, in 1984, Marshall and Warren published a *Lancet* paper detailing their findings. Still, however, physicians refused to acknowledge their theory.

> **IRONICALLY, THE FACT THAT THE DISCOVERY HAD NOT ALREADY BEEN MADE WAS PREVENTING MANY PHYSICIANS AROUND THE WORLD FROM MAKING IT. ATTEMPTING TO EXPLAIN HIS OWN CONVICTION IN THE FACE OF SUCH OPPOSITION, WARREN LATER WROTE, 'I PREFERRED TO BELIEVE MY EYES, NOT THE MEDICAL TEXTBOOKS OR THE MEDICAL FRATERNITY.'**

It was later that year, in 1984, that an impatient and frustrated Marshall undertook his self-experiment. He drank a flask teeming with *Helicobacter pylori*, the organism he and Warren kept finding in the gastric lining of gastritis patients. His efforts were both successful and widely reported. Both gastritis and the bacterium were found thereafter in his stomach, and after a course of antibiotics that eradicated the microbe, his gastritis resolved. They published the result in the *Medical Journal of Australia* in 1985.

But medicine was slow to come around, and it was only in the 1990s that Marshall and Warren's work began to receive serious attention. At that point, physicians and pharmaceutical firms began to develop regimens to eradicate the bacteria, making it possible to remedy what had long been considered a chronic disease that could only be managed but never satisfactorily cured. From that point forward, gastritis and peptic ulcers would be regarded as largely infectious diseases.

Opposite: Model of ranitidine, a medication that blocks gastric acid secretion.
Above: Barry Marshall.
Right: Robin Warren

LEGACY

Marshall and Warren shared the 2005 Nobel Prize in Physiology or Medicine for their discovery of "the bacterium *Helicobacter pylori* and its role in gastritis and peptic ulcer disease". Subsequently, it has emerged that many stomach cancers develop in the context of long-standing gastric inflammation, so eradicating the bacterium can also help to decrease the number of patients who develop cancer.

Today, when patients present with symptoms such as abdominal pain, vomiting, weight loss, and loss of appetite, leading to a suspicion of peptic ulcer disease, physicians can test for the presence of *Helicobacter pylori* using a blood test for antibodies, a breath test for a gas the bacteria produce, or a biopsy from the stomach. Antibiotics play an important role in treatment, though as a student of microbiology might predict, physicians are encountering more and more antibiotic-resistant cases.

30 VACCINATING AGAINST CANCER: HPV

Cervical cancer is both the fourth most common type of cancer and the fourth most common cause of cancer death among women worldwide. By the time patients present with symptoms such as pelvic pain and vaginal bleeding, the disease has often progressed to a point that it cannot be cured. To reduce mortality, it is necessary to catch cancers sooner.

IMMORTAL CELLS

Although she lived in obscurity during her lifetime, Henrietta Lacks (1920–51) was to become one of the best-known patients to die of cervical cancer. A native of Virginia, she led a tough life. Her mother died giving birth when Lacks was four years old, and she and her siblings were raised by relatives. Lacks grew up with her grandfather in a cabin that once served as slave quarters.

She gave birth to her first child when she was 14 years old, and she and her husband later moved to Baltimore. There she was diagnosed with cervical cancer in the last year of her life. She was treated with radiation but developed severe abdominal pain. She required multiple blood transfusions. After her death, an autopsy revealed widespread metastatic disease.

Lacks has become famous because her cancer cells turned out to divide at a very fast rate and to survive much longer than most cells. Because they did not die after many divisions, they were referred to as "immortal". A cell line was developed that soon became known as "HeLa" cells, which were used for experiments in laboratories around the world. For example, Jonas Salk used HeLa cells to test the first polio vaccine.

It has been estimated that as many as 50 million metric tons of the cells may have been produced and that over ten thousand patents have been issued for work performed using them. Permission from Lacks and her family was not obtained, and many ethical questions have been raised.

Above: Henrietta Lacks.
Right: Georgios Papanikolaou.

THE PAP TEST

The first major advance in the reduction of cervical cancer mortality was the Pap test, developed in the 1920s by Greek physician Georgios Papanikolaou (1883–1962). He obtained cells from the cervix and showed that they could be seen under the microscope. When precancerous cells are detected, abnormal tissue can be removed from the cervix, preventing the development of cancer.

While the Pap test helped to reduce cervical cancer mortality by as much as 70 per cent in the US, moving cervical cancer from the number-one cancer killer of US women in the early twentieth century to number 12 today, it does not prevent the early changes that can lead to cancer. To do that, it would be necessary to address the underlying cause of such cancers.

HPV AND CANCER

Harald zur Hausen (b. 1936), a German virologist, developed the hypothesis that cervical cancer is caused by the human papillomavirus (HPV). The association was suggested by numerous reports of genital warts, which were already known to contain papillomaviruses, transforming into cancers. He obtained tissue from numerous wart biopsies.

Zur Hausen's investigations showed that viruses present in some warts were not present in others, suggesting the existence of multiple types of the virus. In the 1980s, he demonstrated that certain types of the virus, HPV-16 and HPV-18, can be found in cervical cancers. For his efforts, zur Hausen was awarded the 2008 Nobel Prize in Physiology or Medicine.

Over one hundred species of papillomavirus have been recognized, and different types have been identified in mammals, birds, reptiles, and fish. They were first identified in the early twentieth century in association with skin warts or papillomas, hence their name. In humans, most HPV infections cause no symptoms and resolve within several years.

In some cases, however, the infection can cause warts or even precancerous lesions, most commonly in the uterine cervix, but other sites are possible. It is estimated that about 70 per cent of cases of cervical cancer are due to HPV-16 and HPV-18, while other types cause genital warts. The types that cause cancer are generally spread by sexual contact, and it is usually acquired shortly after the onset of sexual activity.

In patients who develop cancer, it is thought that about 20 years typically elapse between infection and the development of cancer. However, this period of time can be shortened in patients with impaired immune systems, such as untreated HIV infection. Risk factors for the development of cancer include other sexually transmitted infections, such as herpes, chlamydia, and gonorrhoea.

DEVELOPING A VACCINE

Epidemiologists, virologists, and physicians worldwide collaborated to develop vaccines against the types of HPV that most commonly cause cervical cancer. Much of the work was performed at the University of Queensland in Australia, although multiple US universities and institutions also played a major role in developing the vaccine's final form.

The vaccine uses virus-like particles that resemble specific types of HPV. However, these particles are hollow and contain no viral genetic material, meaning that they cannot cause infection or increase cancer risk. However, they are sufficiently similar to HPV that they evoke an antibody-mediated immune response, which then attacks the virus if vaccinated individuals are exposed to it.

Such vaccines are virtually 100 per cent effective in preventing infection, and if patients cannot become infected with HPV, they will not develop HPV-related cancers. Moreover, the vaccines appear to cause no significant side effects and appear to be essentially free of significant risks. It is recommended that girls receive the vaccine at puberty.

Left: Harald zur Hausen.
Below: The HPV vaccine.

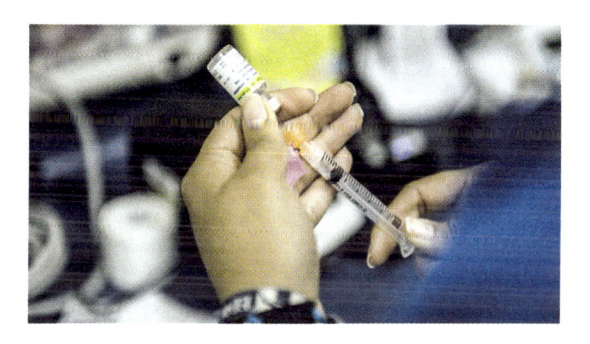

31 INFECTIOUS DISEASE AS A WEAPON: BIOTERRORISM

It has long been known that infectious diseases can be used as weapons. In ancient times, arrows were contaminated with tissue from dead bodies, wells were contaminated with poisonous plants and the bodies of dead animals, and poisonous snakes were hurled onto the decks of enemy ships. More recently, blankets contaminated with smallpox were given to native populations, fields around cities were flooded to promote malaria, and major world powers developed offensive biological weapons programmes.

WHAT IS BIOTERRORISM?

Bioterrorism involves the deliberate use or threat of use of a biological agent. One feature of such an attack is its targets. The most obvious target would be humans, who can be infected with agents that cause death or incapacitation. But bioterrorist attacks may also be directed at other targets. For example, food supplies can be disrupted by attacks on plants and animals, thereby damaging both economic capacity and morale.

Another parameter of bioterrorism is the medium of attack. For example, infectious agents and toxins can be distributed through the air by dropping them from aircraft, dispersing them with explosions, or releasing them in contained spaces, such as a subway. Another medium is water, the effects of which can be amplified by targeting central distribution points. Finally, food can be an effective means of dissemination, especially attacks on food-processing facilities.

Another key feature of a bioterrorist attack is the agent itself. These include bacteria, viruses, insects, fungi, and toxins. More specific examples include attempts to infect large numbers of people with the bacterium that causes anthrax, the use of the smallpox virus, to which fewer and fewer people have been exposed, the dissemination of plague using infected fleas, the contamination of food supplies with fungi, and the use of neurotoxins such as botulism toxin.

Some bioterrorist agents are transmissible from person to person and others are not. For example, anthrax spores must generally be inhaled, ingested, or contacted through broken skin, and do not spread easily from person to person. By contrast, a viral infection such as smallpox is easily transmissible through the respiratory system, and infecting just a few people could rapidly produce an epidemic or even worldwide pandemic.

Motivations for bioterrorism may also vary. In a military context, bioterrorism may inflict as much damage on an opposing force as more conventional weaponry such as firearms, artillery, and explosives. Moreover, such an attack can damage personnel without destroying supplies, equipment, and buildings. In a civilian setting, bioterrorism may be used to change government policy, or simply to coerce or intimidate a civilian population.

Some biologic agents can be difficult to control. Suppose, for example, that an agent is being spread through the air. A mere shift in winds could redirect such an agent away from enemy forces and toward the side deploying it. Likewise, a highly infectious agent

Opposite:
The smallpox virus.

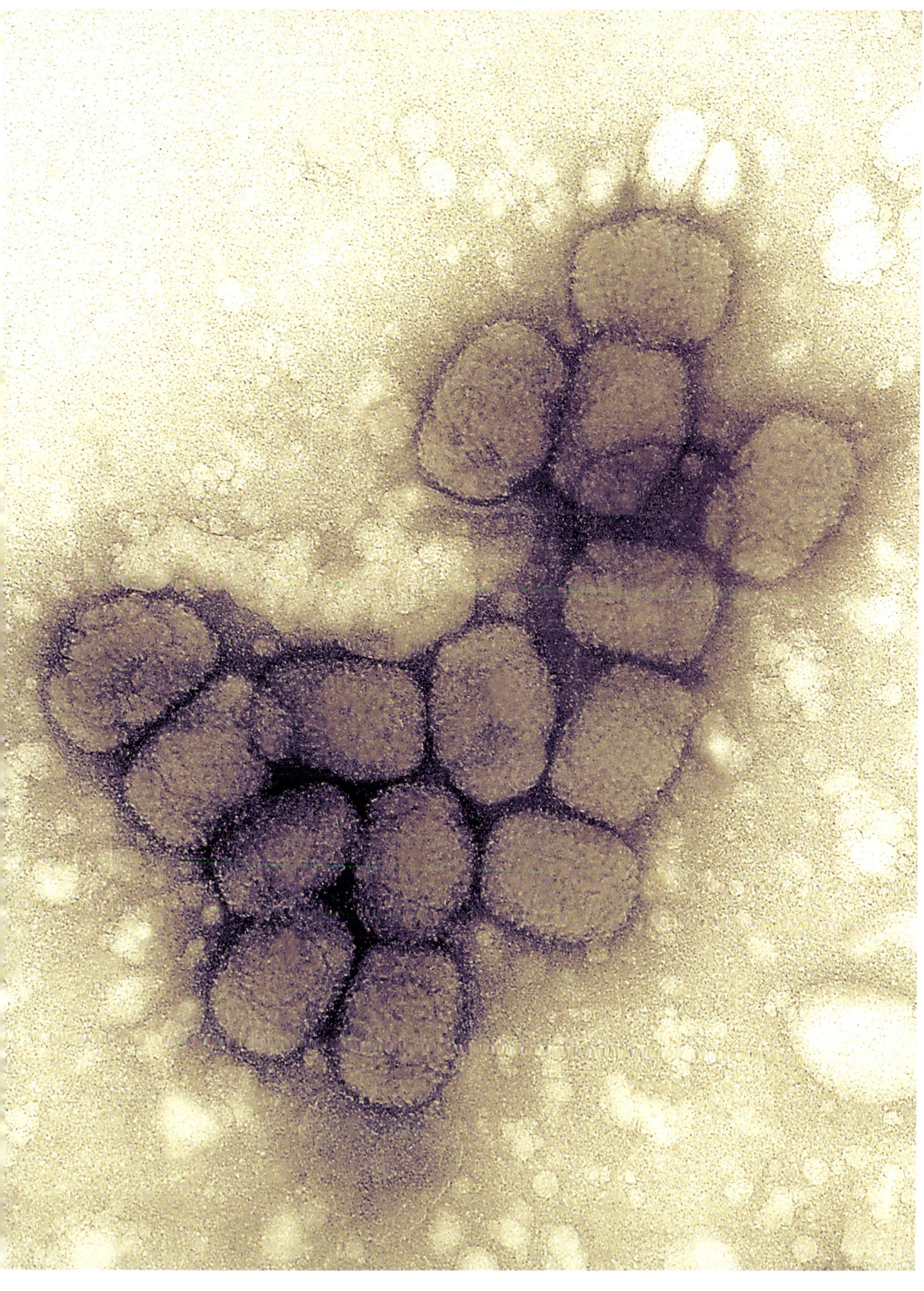

such as smallpox might be released in one population but quickly spread across borders to infect neutral or even friendly forces. Bullets and bombs, by contrast, can generally be more precisely controlled.

AN ATTACK IN NEW YORK

One example of a bioterrorist attack occurred in the US in 2001. In October of that year, a newspaper editor became infected with anthrax. Soon, two of his colleagues fell ill, and an investigation disclosed anthrax spores throughout much of the building in which they worked. Other cases appeared in New York media companies. It turned out that victims had come into contact with letters containing a white powder. Before long, such letters started to appear in US congressional offices.

By the end of the following month, no more letters were found, but nearly two dozen people had developed anthrax infections, causing five deaths. While the number of infected individuals was low, media attention and public concern were high. It is estimated that in the following months over 100,000 samples were tested, overburdening laboratories and public health authorities; tens of thousands of people were treated with antibiotics; and many hoaxes and false alarms added to the confusion.

The public alarm over the attacks was magnified by the fact that they unfolded in the weeks after the September 11 terrorist attacks. Various suspects were identified, and one, a scientist at the US government's biodefence laboratories, committed suicide after being placed under government surveillance. Some speculated that the alleged attacker stood to benefit financially from the attacks because he had helped to develop two anthrax vaccines.

POTENTIAL DEVELOPMENTS

Of course, those seeking to weaponize biologic agents are not necessarily restricted to existing pathogens. New biotechnologies could be employed to design novel agents. For example, microbes that currently do not cause disease might be transformed into virulent forms, or the virulence of existing microbes might be amplified. Also, new types of bacteria or viruses might be created that are resistant to existing therapies, such as antibiotics.

Microbes might also be rendered more transmissible. For example, the microbe responsible for bubonic plague might be made more transmissible from person to person by a respiratory route. Still another means of increasing the damage that a biologic agent could inflict would be to render it less detectable. For example, if the period during which an infection remains asymptomatic could be lengthened, it might spread much more widely in a population before it came to attention.

Infectious Disease as a Weapon: Bioterrorism

BIOSURVEILLANCE

The detectability of bioweapons naturally raises the issue of biosurveillance. What systems do societies have in place to detect bioterrorist attacks, and what systems need to be developed? One source of such information comes from the healthcare system, such as healthcare records in physician offices, clinical laboratories, and hospitals. But other diverse sources, such as internet search patterns and veterinary health records, might also prove crucial.

The fact that many such records are now electronic has enhanced the ability to monitor them, and public health specialists are on the lookout for spikes in symptoms that could represent bioterrorist activity. To counteract such attacks, it is important that they be detected as early as possible, before an infection has spread widely in a population. Many nations have specialized military units that can be deployed if such an attack is detected.

Opposite, from top: Anthrax reward poster showing envelopes that contained anthrax; one of the anthrax letters.
Above: The scene of a sarin gas attack in the Tokyo subway.

32 CORONAVIRUSES: TWENTY-FIRST-CENTURY PANDEMIC KINGS

Named for their crown-like appearance on electronic microscopy, the coronaviruses were first discovered in chicks in the 1930s and recognized as a cause of human disease in the 1960s. They are RNA viruses whose clinical manifestations in patients range from asymptomatic to the common cold to a lethal disease.

The most common mode of transmission is by means of microbes moving through the air, which are produced by speaking or coughing and then inhaled by another person. The spikes on the surface of the virus, which account for its corona-like appearance, serve to attach the virus to receptors on host cells, which are usually the ones lining the respiratory tract.

Coronaviruses have been around for at least thousands and likely millions of years, and bats and birds are thought to constitute key reservoirs. There are multiple species of coronavirus, some of which are thought to cause about 10 to 20 per cent of common colds. In recent years, however, three have been responsible for sometimes fatal disease: SARS-CoV, MERS-CoV, and SARS-CoV-2.

SARS stands for Severe Acute Respiratory Syndrome, MERS for Middle East Respiratory Syndrome, and CoV for coronavirus.

SARS

In late 2002, an outbreak of a previously unknown coronavirus infection now known as SARS-CoV appeared in China's Guangdong province, near Hong Kong. In January of 2003, a fishmonger was admitted to a hospital there, where more than two dozen health professionals were infected. In February, a hospital staff

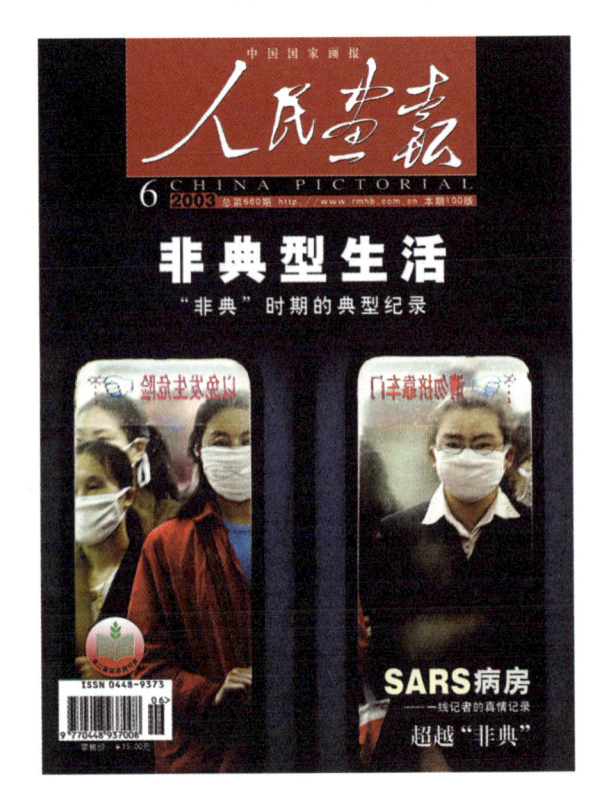

Above: Chinese poster showing people wearing masks during the 2003 SARS outbreak.

Opposite, from top: During the SARS outbreak, extra care had to be taken to disinfect public spaces; electron micrograph of the SARS-CoV-2 virus.

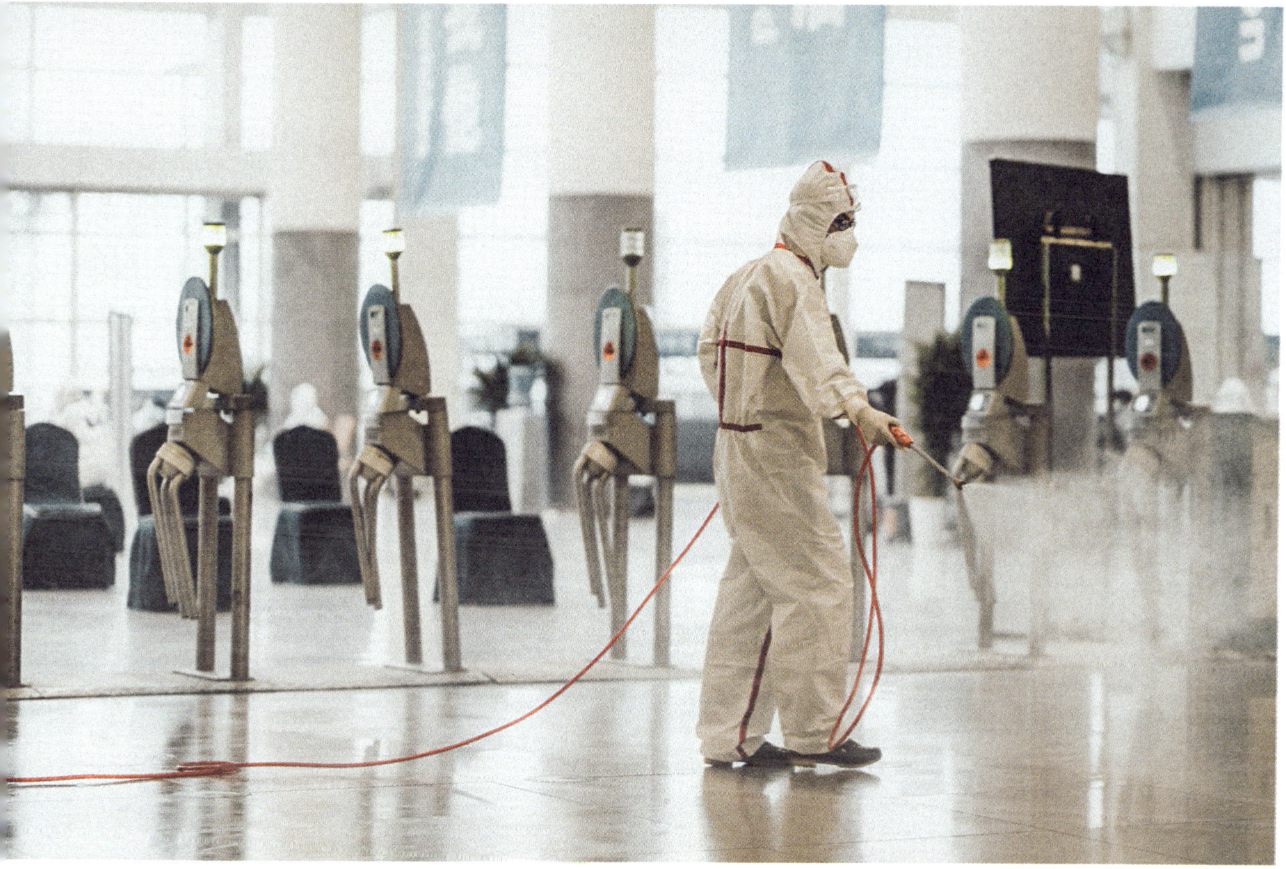

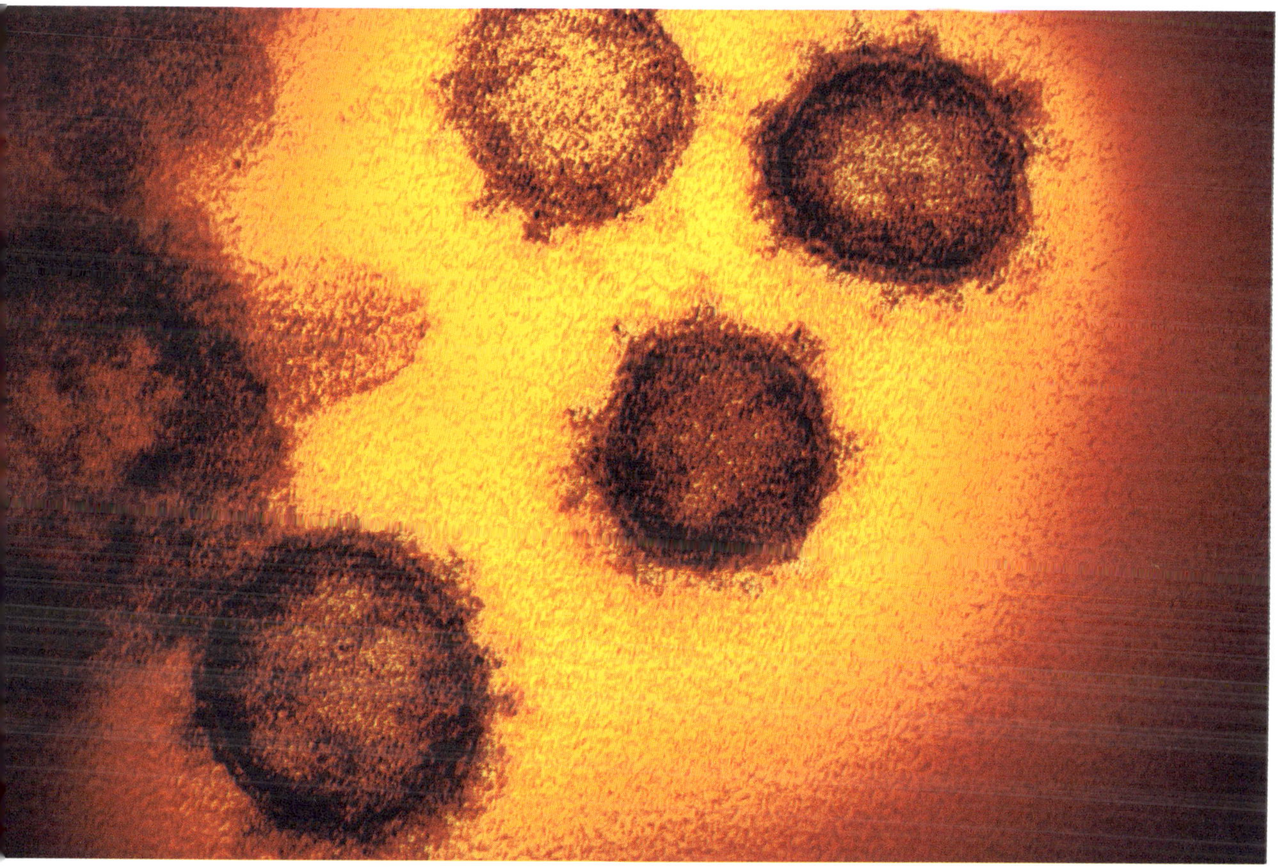

member attended a wedding in Hong Kong, where many hotel guests were infected. Travellers brought the disease to Hanoi and Toronto, and it soon spread elsewhere, reaching the US and Europe.

The outbreak was declared contained by the World Health Organization in July of 2003, with the last cases reported in May of that year. In total, it is estimated that over eight thousand people were infected, with nearly eight hundred deaths. Healthcare workers bore a large portion of the disease burden, accounting for about one-fifth of cases around the world, including over forty deaths, especially among physicians and nurses.

Standard preventive measures against coronavirus infection include hand hygiene, disinfection, and isolation. Hospitalized patients are often kept in rooms with negative pressure, and patients with severe illness usually require mechanical ventilation. It appears likely that much of the severe damage associated with coronavirus infections is due to an excessive response of the host immune system. In part because no cases of SARS-CoV have been detected since 2003, no vaccine or specific antiviral therapy was ever developed.

MERS-COV

The first cases of MERS were identified on the Arabian Peninsula in 2012, and a total of between 2,500 and 3,000 have been seen so far. Although this coronavirus likely originated from bats, most human infections are thought to arise from camels. Because human-to-human transmission, at least outside of hospitals, is infrequent, the threat of a worldwide MERS pandemic is low. However, this is the most lethal of the human coronavirus infections, causing death in up to one-third of patients.

Patients manifest typical symptoms of a coronavirus infection, including fever, cough, shortness of breath, and muscle aches and pains. A minority of patients complain of gastrointestinal symptoms, including diarrhoea, vomiting, and abdominal pain. Some MERS patients are asymptomatic, and there appears to be essentially no risk of disease transmission by asymptomatic patients. MERS has been seen in Europe and the US, for example in a healthcare worker who had been in Saudi Arabia a week prior to diagnosis.

> **CORONAVIRUSES HAVE BEEN AROUND FOR AT LEAST THOUSANDS AND LIKELY MILLIONS OF YEARS**

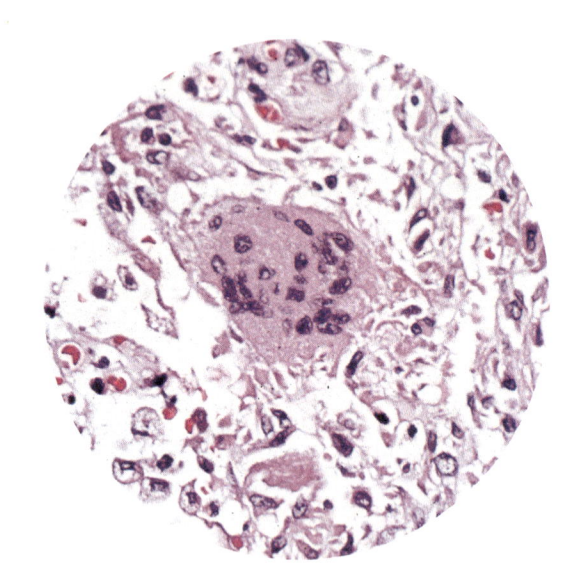

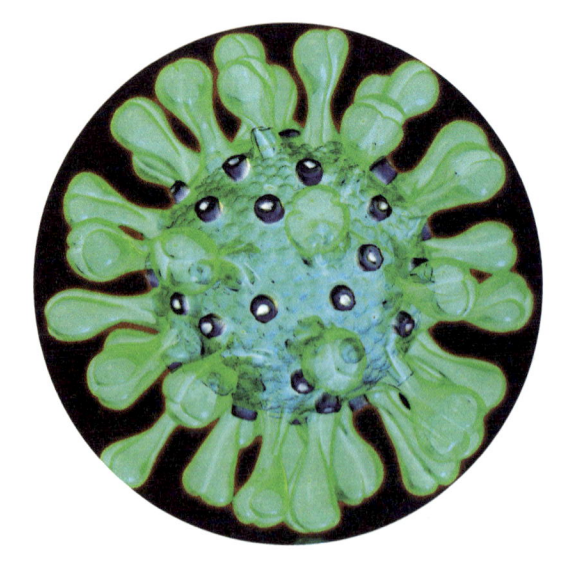

Above, from top: Micrograph showing changes to lung tissue due to SARS; illustration of the MERS coronavirus.
Opposite: A makeshift COVID-19 hospital in Madrid.

SARS-COV-2

COVID-19, or Coronavirus Disease 2019, was first identified in Wuhan, China, in late 2019 and spread around the world in a matter of a few months. The infection is thought to have originated in bats, but unlike SARS-CoV and MERS, this virus is particularly transmissible from human to human. Some have speculated that the first cases arose in connection with a live animal market in Wuhan, while others have suggested that the point of origin may have been a research laboratory.

The fatality rate of COVID-19 is not nearly as high as MERS – perhaps just 1 per cent or lower – but because the disease is so highly transmissible, with 2 million cases worldwide by late April 2020, it causes far more deaths. In general, the lethality of the disease increases with patient age, with many children remaining asymptomatic. Of interest, more men than women appear to die from the disease, and chronic respiratory illnesses and heart disease also increase the risk of death.

One of the most notable cases of COVID-19 infection was that of a Chinese physician who was among the first to draw attention to the new disease. Li Wenliang was born in Beizhen in 1986. He attended Wuhan University School of Medicine, during which he joined the Communist Party. After graduation, he trained in ophthalmology and in 2014 he began practising at Wuhan Central Hospital.

In late 2019, Li encountered a laboratory report indicating that a patient had tested positive for a SARS coronavirus. He shared the result via social media with some of his colleagues, indicating that multiple patients had tested positive for the virus.

When his message, which included a CT-scan image of a patient's lungs, were shared on the internet, he was reportedly summoned by his hospital's leadership, where he was chastened for unauthorized leaking of information. He was persuaded to sign a statement in which he admitted to disturbing public order. In January of 2020, after signing the statement, Li returned to work, but as word of the outbreak spread, he was accused along with other individuals of "rumour mongering".

STAY HOME
PROTECT THE NHS
SAVE LIVES

国际出发 International Departures
(港澳台)
(HongKong-Macau-Taiwan)
计划时间 Scheduled
航班号 Flight
终点站/经停站 To/Via
区 Area
航班状态 Status
17:30 CA163 / NH5726 大阪 OSAKA H 取消 CANCELLED
17:50 AC026 / CA7455 温哥华 VANCOUVER 取消 CANCELLED
18:05 AA128 达拉斯 DALLAS 取消 CANCELLED
18:15 AC012 / CA7519 蒙特利尔 MONTREAL 取消 CANCELLED
18:15 AC028 / CA7457 多伦多 TORONTO 取消 CANCELLED
18:20 AA182 洛杉矶 LOS ANGELES 取消 CANCELLED
香港 HONG KONG
大阪 OSAKA
香港 HONG KONG
高雄 KAOHSIUNG
洛杉矶 LOS ANGELES
13:55 IJ002 TOKYO/NARITA
14:15 NZ288 奥克兰 AUCKLAND
14:45 KA805 / CX5805 香港 HONG KONG
15:25 VN523 / MU8474
15:25 VN531 / MU8472 河内 HANOI
15:40 KA893 / CX5893 香港 HONG KONG
16:05 MH389
16:25 OZ366
16:25 SQ833 / VA5551
16:30 UA087 / CA7215

Li later complained that he had been chastised for telling the truth. The next month, the Chinese judiciary issued a statement arguing that Li and other individuals should not have been punished for their claims. In fact, said the court, "It might have been a fortunate thing if the public had believed the 'rumours' then and started to wear masks, carry out sanitization measures, and avoid the wild animal market" where the outbreak may have originated.

Li later stated that he had been unfairly criticized, saying, "I think there should be more than one voice in a healthy society, and I don't approve of using public power for excessive interference." Li, however, fell ill himself, likely after treating a patient with COVID-19, and he was admitted to intensive care in January 2020. He died in February, but not before he posted a message on social media declaring his intention to return to his medical practice after he recovered.

Both Li's parents were infected with the coronavirus but recovered. As soon as Li fell ill, he moved into a hotel, in an effort to protect his pregnant wife and their son from contracting the disease. A faculty member at Peking University said of Li:

I deeply mourn for all the medical practitioners passing away in the struggle against this emerging infectious disease, especially Dr Li Wenliang, as one of the whistle-blowers dedicating his young life in the front line. We were encouraged by his dedication to patients and we will continue to fight against the virus to comfort the dead with the final victory.

Regarding Li's contribution, a faculty member at Johns Hopkins University said:

One of the world's most important warning systems for a deadly new outbreak is a doctor's or nurse's recognition that some new disease is emerging and then sounding the alarm. It takes intelligence and courage to step up and say something like that, even in the best of circumstances. Rising doctors and nurses should remember Dr. Li's name for doing the right and brave thing for his community and the world, and should be encouraged to do the same if they are even in a moment to make that kind of difference in the world.

LESSONS TO LEARN

The coronaviruses serve as a timely reminder that potentially pandemic diseases are ever present somewhere in the world, and all that is required for the next outbreak to occur is a chance mutation in a microbial genome or an encounter with a reservoir species, such as a bat or a bird. One of the key steps in preventing pandemics is to reduce known sites of transmission such as markets selling live wild animals and to detect outbreaks of new infections as early as possible.

Opposite, from top: An empty street in Leeds during the COVID-19 pandemic, a sign that people are taking social distancing seriously; a Hong Kong shopper in a mask next to empty supermarket shelves during the early days of the COVID-19 pandemic; a worker in protective gear gazes up at an airport screen in Shanghai during the COVID-19 pandemic.
Below: A symptomatic patient has her temperature checked in Wuhan, China in February of 2020.

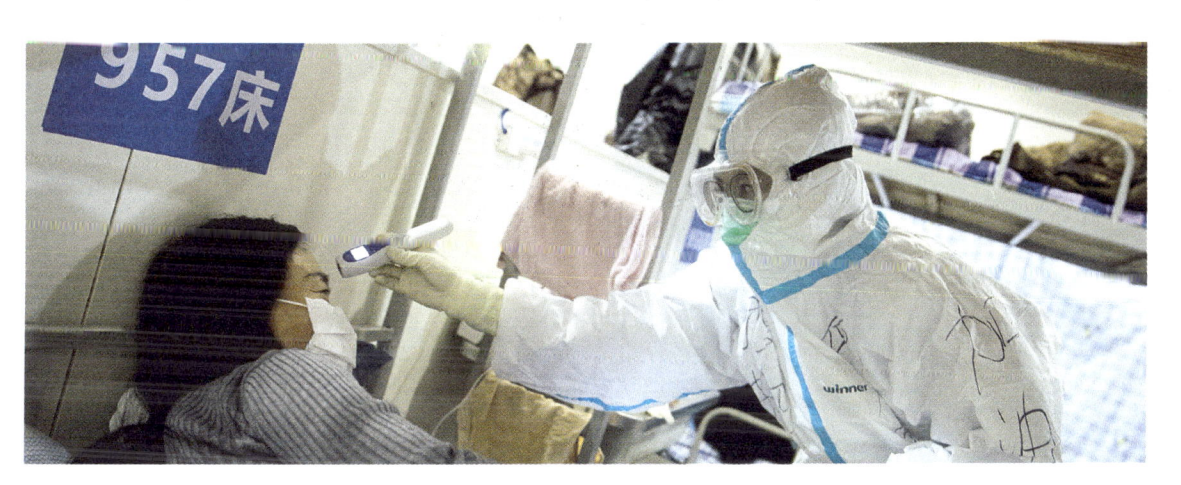

33 INFECTIOUS DISEASE: THE ROAD FORWARD

The philosopher George Santayana (1863–1952) wrote that those who cannot remember the past are condemned to repeat it, to which there stands a corollary: those who know history can avoid replicating the same mistakes. Immersing ourselves in the history of infectious disease offers numerous insights that can and should shape how individuals, families, communities, nations, and humanity move forward in the future.

A SHRINKING WORLD

One such lesson concerns the size of the world itself. While the planet earth remains a huge place, in other respects it has been shrinking rapidly. For example, the human population in 1900 is thought to have been about 1.6 billion, while today it is moving rapidly toward 8 billion and may reach 10 billion in just 60 more years. This raises concerns about resources such as food and water, but it also means that the earth is becoming more crowded, which promotes transmission of many microbes.

The growth in the absolute number of human beings tells only part of the story. Not only are there more people than ever before, but the proportion of people living in relatively crowded conditions is also increasing. The United Nations has estimated that 2007 was the first year when more humans lived in urban than rural areas, and by 2050 it is projected that two-thirds of people will be living in cities, where higher population densities further promote the spread of infectious disease.

Even this does not tell the whole story. In relatively rich countries, as much as 80 per cent of the population often lives in urban areas. In this sense, being well-off does not necessarily offer protection against the spread of epidemic disease. On the other hand, about one in three people who live in cities dwell in slums – defined as habitations with poor water or sanitation, little living area, and poor housing durability. For these reasons, slums provide almost ideal conditions for the spread of many infectious diseases.

THE BUTTERFLY EFFECT

A second lesson concerns what is sometimes referred to as the butterfly effect: the idea that large effects can flow from surprisingly small and remote causes. Originally developed in meteorology, the idea has been colloquially illustrated by the example of the path of a tornado in the United States that has been influenced by the flapping of a butterfly's wings several weeks earlier somewhere in South America. Double pendulums (where one pendulum is attached to the end of the other) provide very direct examples of this effect, moving in a chaotic manner.

How does the butterfly effect apply to infectious disease? Consider HIV/AIDS,

Above: Artist's image of a butterfly.
Opposite: Europe and North Africa at night, showing major air routes.

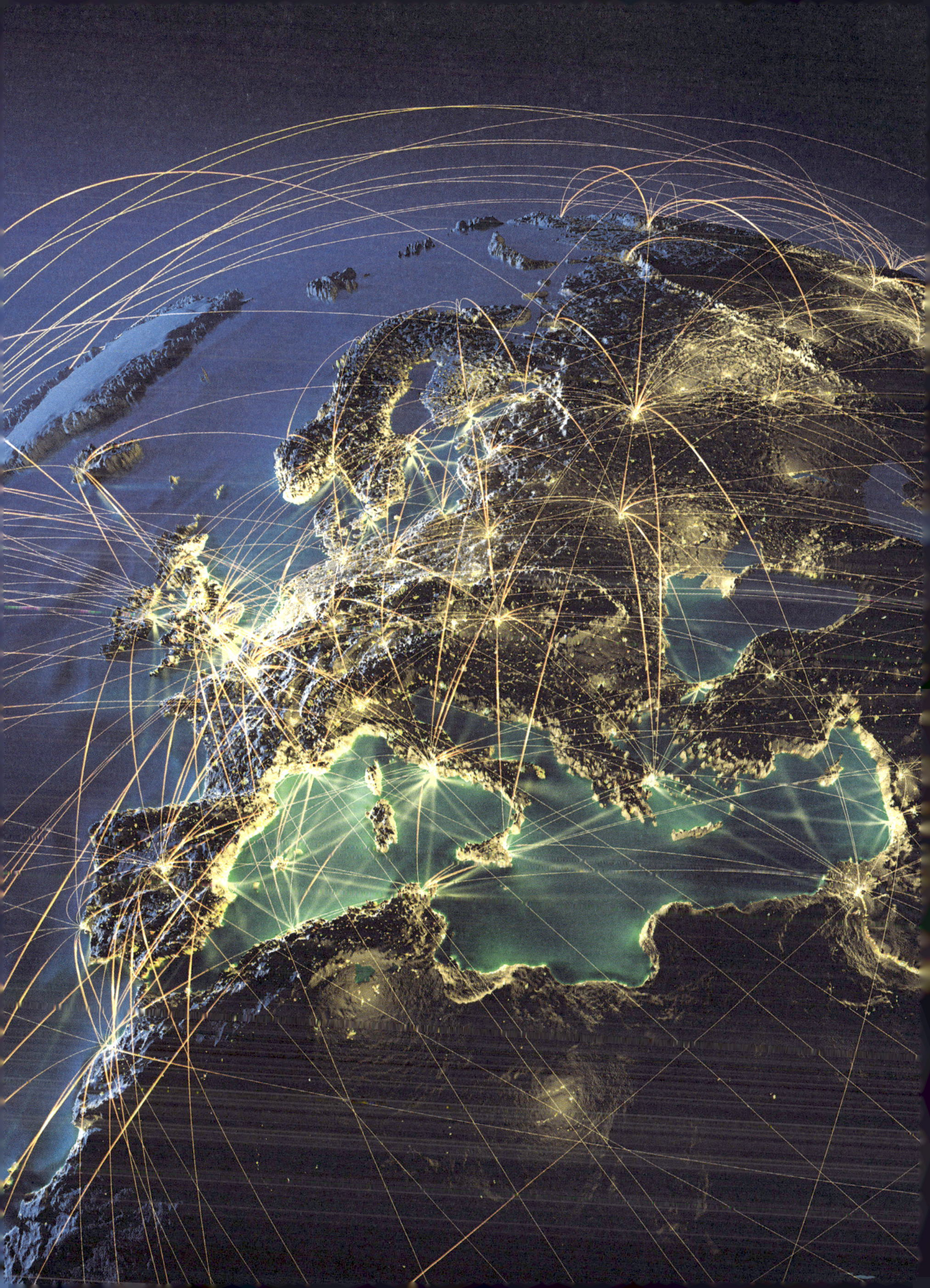

a disease that has killed tens of millions of people worldwide, with many more infected by the virus. Recent genetic studies have suggested that the forerunner of HIV was first transmitted to human beings in the Democratic Republic of Congo in the 1920s. Had a particular hunter not eaten a particular chimpanzee or been infected by a particular chimpanzee's blood, the worldwide pandemic might have been averted.

Ebola haemorrhagic fever, one of the deadliest of viral diseases, is thought to have existed for a long time in bats, which can spread it to other animals such as apes and monkeys. In 1976, the first human beings we know of became infected in South Sudan, resulting in the deaths of several hundred people. Several dozen outbreaks have occurred since. Once the virus makes the jump to humans, it can be spread by contact with blood and other bodily fluids.

A third more recent example of the butterfly effect is the origin of the worldwide COVID-19 pandemic. While the jury is still out, it appears likely that the coronavirus originated in bat populations and may have made the jump from an intermediate animal species to humans in a wet market – where live animals are traded – in Wuhan, the capital city of Hubei province in the People's Republic of China. Again, had a single interaction not occurred, it is possible that hundreds of thousands of lives would not have been lost.

One implication of such outbreaks is the value of disease surveillance and early intervention. When only a small event can produce such dramatic and widespread consequences, preventing such events and intervening early in the causal chain can offer huge benefits, containing epidemics and pandemics while they are still confined to relatively few people. As such infections spread, the harm they cause and the resources necessary to counteract them grow exponentially.

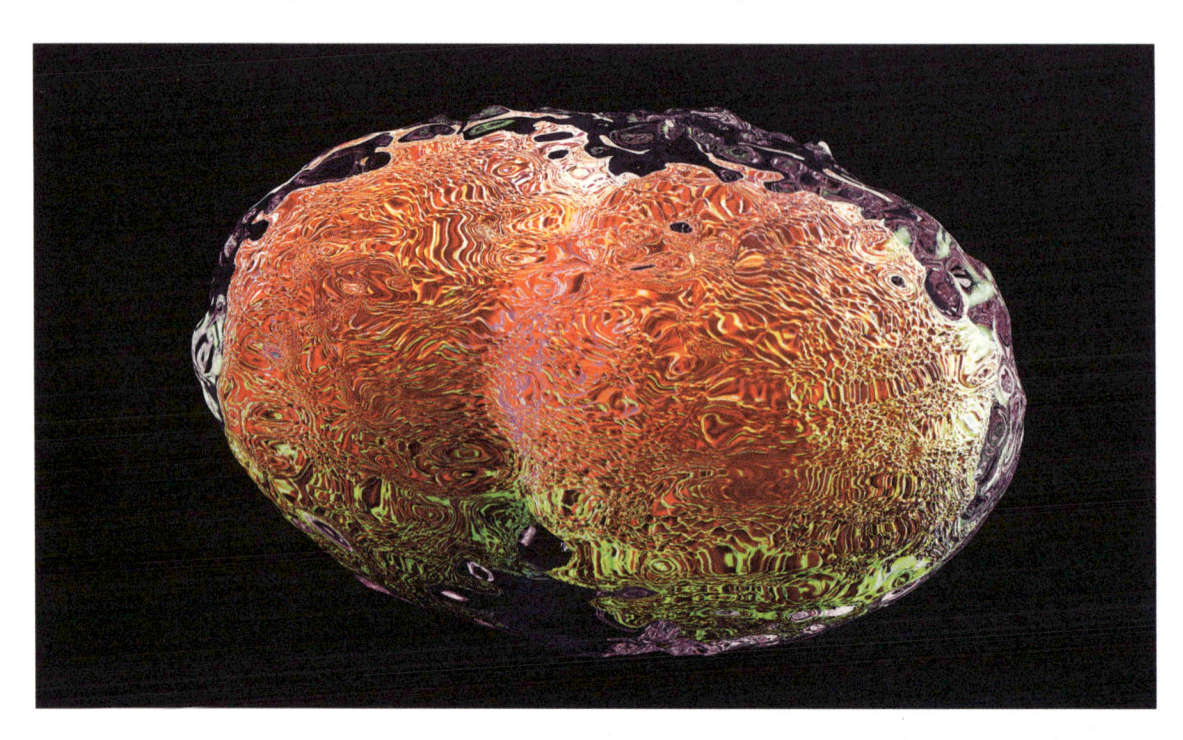

Left: Kent Brantly, US physician and former student of the author's, after treatment for Ebola virus infection acquired while in Liberia.
Below: An illustration of the smallpox virus.
Opposite: A panel of tests for respiratory viruses being used in France during the COVID-19 pandemic.

> **THERE WILL BE MICROBES AS LONG AS THERE ARE HUMAN BEINGS … OUR MISSION IS NOT TO VANQUISH THEM, BUT TO UNDERSTAND AND LEARN TO COEXIST AND EVEN THRIVE WITH THEM.**

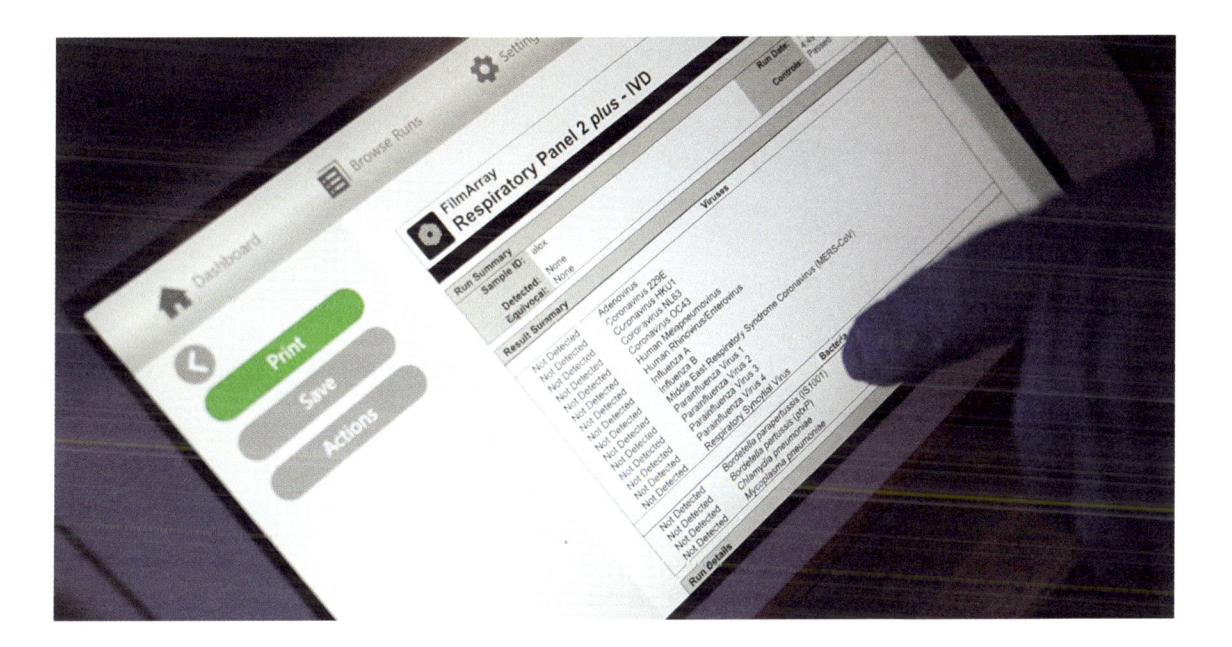

BEYOND BORDERS

A third insight from the history of infectious disease is the relative unimportance of national boundaries. While human beings may regard such borders as impenetrable walls, they are usually imperceptible to other species. Birds, bats, and monkeys often cross them freely, and the same goes for bacteria, viruses, fungi, parasites, and the other organisms that cause infectious diseases.

Ironically, one predictable consequence of the COVID-19 epidemic has been the closure of national borders to many travellers. Yet when governments focus too much on protecting their own citizens, they can overlook opportunities for collaboration and solidarity that reflect the fact that we are all "citizens" of one world. If we respond to a global challenge such as a pandemic with a strictly nationalist, us-versus-them mentality, we may end up making the situation worse.

A BIOLOGICAL COMMUNITY

This idea of global citizenship can and should be expanded even further. For we are not just citizens of a human community. We are also citizens of a biological community, a worldwide community of living organisms that includes plants, animals, fungi, protists, bacteria, archaea, and even viruses of almost unimaginable diversity and richness. In one respect, each is competing with the others, but from another perspective, they are mutually dependent on one another.

To survive and thrive in such a complex biological nexus, we need to see ourselves less as standing in mastery above it all, and more as neighbours and perhaps even collaborators. There will be microbes as long as there are human beings – in fact, we cannot survive without them. And should humanity ever exit the stage of life on earth, they would almost certainly long outlive us. Our mission is not to vanquish them, but to understand and learn to coexist and even thrive with them.

FURTHER READING

American Museum of Natural History. Epidemic!: The World of Infectious Diseases. New York: The New Press, 1999.

Barry, John. *The Great Influenza: The Story of the Deadliest Pandemic in History*. New York: Penguin Books, 2005.

Bolker, Benjamin and Wayne, Marta. *Infectious Disease: A Very Short Introduction*. New York: Oxford University Press, 2015.

Camus, Albert. *The Plague*. New York: Vintage, 1991.

Defoe, Daniel. A Journal of the Plague Year. New York: Penguin, 2003.

Garrett, Laurie. *The Coming Plague: Newly Emerging Diseases in a World Out of Balance*. New York: Penguin, 1995.

Kasper, Dennis and Fauci, Anthony. *Harrison's Infectious Diseases*, 3rd ed. New York: McGraw-Hill, 2016.

McKeown, Thomas. *The Origins of Human Disease*. New York: Oxford, 1988.

McNeill, William. *Plagues and Peoples*. New York: Norton, 1976.

Oshinsky, David. *Polio: An American Story*. New York: Oxford University Press, 2006.

Rosen, George. *A History of Public Health*. Baltimore: Johns Hopkins Press, 2015.

Saramago, José. *Blindness*. New York: Harcourt, 1998.

INDEX

A
Abraham, Edward 124
Acquired Immune Deficiency Syndrome see HIV/AIDS
Adenovirus 8
agar 100
anaesthetics, inhalation 80
anatomic localization of disease 27, *27*
animalcules 51
anthrax 94, 96, 97, 98, 100, 101, 142, 144, *145*
Anthrax bacilli 98
antibiotic resistance 19, 22
antibiotics 19, 22, 120, 121
antibodies 16
antiseptic sprayer 86
Arabian Peninsula 148
artemisinins 116
arthropods 10
aspirin poisoning 107
Athens 28, 29
autopsies 84
Aztecs 44, 45, 46, 47

B
Bacillus 73, 74
bacteria 10, 13, 14, 15, 35, 51, 74, 84, *84*, 96, 98–9, *98*
  "Bad" 26
  comma-shaped 100–01
  rod-shaped 98, 100
  "Good" 26
bacterial culture 100
Barbados 54
Barré-Sinoussi, Françoise 130
BCG vaccine 73
beta-lactam ring 124
beta-lactamase 124
biological agents 142, 144
bioterrorism 142–5
Black Death 32–7, *32*, 38, 40, *41*
  spread 35, *37*
bleeding 62
blood 25, *27*
blue bile 25, *27*
Boccaccio, Giovanni 38, *38*, 41
bones 70
Boston 55
Boyle, Robert 53
brains 70
Brantly, Kent *154*
Brawne, Fanny 77, 78, *78*
Breck, Samuel 62
British army 56
Broad Street pump 82–3, *83*, 126
bronchitis 8
bubonic plague 10, 32, *35*, 36, 126, 144
burial 41
butterfly effect 152, *152*
Byron, Lord 76

C
Candida albicans 10
Canterbury Tales 38
carbolic acid see phenol
Carson, Rachel 115
cell culture 13, 98
Centers for Disease Control 130
cervical cancer 140–41
Chain, Ernst 124, 125, *125*
Chaucer, Geoffrey 38
chest radiograph 73
chickenpox 16
childbed fever 84–5, *84, 85*
chlamydia 121
Chlamydia trachomatis 121, *121*
chlorine 84
chloroform 80
chloroquine 116
cholera 6, 16, 80, 98, 100–01, 102–3, 126, 136
Cholera bacilli 101
Clark, Lieutenant William 50, 59, 62
Clostridium difficile 26, 27, *27*
cold, common 16
"Common Sense" 58
Concerning Famous Women 38
consumption 100, see also tuberculosis
contagion 13, 73
contaminated drinking water 80, *81*, 82
Continental Army 57, 58
Continental Congress 56, 58
coronavirus 16, 146–51
Cortés, Hernán 44, *44*, 46
COVID-19 62, 63, *148*, 149–51, *149, 151*, 154, *154*, 155
cowpox 65, 67, *96*
Crimean War 90, 92
crusades 45
cryptosporidiosis 8
cultures 138
cytokine storm 60, 108

D
Dance of Death 36
Danse Macabre 36
Darwin, Charles 18, *19*
DDT 115
Decameron, The 38, *38*, 40, 41, *41*
Declaration of Independence 58
Democratic Republic of Congo 154
diagnosis 13
diarrhoeal diseases 8
dissection 25
DNA 132, 133

E
Ebola hemorrhagic fever 154, *154*
Edinburgh, University of 58
Edward VII, King 89
electron micrograph *146*
electron microscopes 53
endotoxins 16
engrafting see inoculation
Epidemiological Society 80
epidemiology 80
ethambutol 73
ether 80
Eukaryotes 10
exotoxins 16
extremophiles 14

F
facemask *109*
Finlay, Carlos 113
First World War 104, 107
fleas 12, 32, *32*, 35
Fleming, Alexander 19, *19*, 20, 122–5, *125*, 126
flights at night *152*
Florence 38, 40
Florey, Howard 124, 125
Franklin, Ben 57
Franklin, Rosalind 126
fungi 20

G
Galileo 50, *50*
gastritis 136, 138, 139
Gay-Related Immune Deficiency (GRID) 130
genetic mutations 19
genital warts 141
gonococcus bacterium 120
gonorrhoea 120–21, *121*, 125
gout 124
Greek healing god *31*
Guandong Province 146
Guatemala syphilis study 120
Guy's Hospital 76

H
haemophiliacs 130
haemoptysis 77
handwashing 84–5, 90, 109

healing god, Greek *31*
Heatley, Norman 124
HeLa cells 140
Helicobacter pylori 136, 139
herpes 121
Herpes simplex virus 121
highly active antiretroviral therapy (HAART) 134
Hippocrates 24–7, *24, 25*
Hippocratic medicine 25, 26, *29*
Hippocratic Oath 24
history of medicine 24
HIV/AIDS 8, 16, 73, 96, 121, 130–35, *130, 133*, 152, 154
   spread *132*
   therapies 134, *134*
Hodgkin, Dorothy 124, 125, *125*
holism 24, 25
Hong Kong 151
Hooke, Robert *12*, 50, 51, 53, *53*
hookworm 126
hospital ward *106*
HPV virus 141
Human Immunodeficiency Virus *see* HIV/AIDS
human papillomavirus (HPV) 141, *141*
humoral theory 25–6, 62
humours 25–6, *27*
Hunter, John 67
hydrochloric acid (HCl) 136
hydroxychloroquine 62, 63
hygiene, poor 90

I

immunization 108, 109
Incas 46
Independence Hall, Philadelphia *59*
industrial revolution 70
infection 13
infectious microbes 16
influenza 16, 37
   A 8, *109*
   B 8
   H3N2 *15*
inhalation anaesthetics 80
inoculation 56, 64, 96, 98
insect repellent 115

insecticides 115
intestinal tuberculosis 100
iron 16
isoniazid 73

J

Jefferson, Thomas 58
Jenner, Edward 56, 64–9, *65, 67*, 96
John Snow Society 83
Johns Hopkins University 151

K

Kaposi's sarcoma 130
Keats, George 76
Keats, John 76, *76*
Keats, Tom 76–7
Koch, Robert 6, 7, 70, 83, 98–101, *98*, 102, 103, 136, 138

L

Lacks, Henrietta 140, *140*
"lady with the lamp" 90
Lake Texcoco 45
lancet *64, 68*
Lancet, The 86, 138
Landsteiner, Karl 126
Laveran, Charles 113
lawlessness 40
leeches *38*
Leeds 151
Leeuwenhoek, Antonie van 50–53, *50*
Lewis and Clark expedition *59*
Lewis, Captain Meriwether 58, *59*, 62
Li Wenliang, Dr 149, 151
light microscopes 53
Lister, Agnes 86
Lister, Joseph 86–9, *86, 89*
   Lister Medal *89*
Listeria 89
Listerine 89
London 35, *35*
lungs 70
lymph nodes 32
lysozyme 122

M

macrophages 70, 73

malaria 6, 7, 8, 22, 113–7, *114, 115*, 126
   distribution 23
   parasite life cycle 115
   symptoms 113
   worldwide *116*
malnutrition 37, 70
March of Dimes 7, 128
Marshall, Barry 136–9, *139*
Mayans 46
measles 8, 46
mechanical respirator *129*
Medical Inquiries and Observations Upon the Diseases of the Mind 58
Medical Journal of Australia 139
medicine, history of 24
mental illness 58, 60, 63
Merck 124
mercury compounds 121
mercury medications 62, 63
MERS 146, 148, *148*
Mexico 44, 45
microbes 16, 19, 22, 50, 94, 144, 146, 155
microbiology 50
Micrographia 12, 50, *53*
microorganisms 86, 89, 100
microscopes 13, 50
microscopy 50, 53, 98, 100
Middle East Respiratory Syndrome *see* MERS
Montagnier, Luc 130
Montague, Lady Mary 64, *67*
Montezuma 44, *45*
Morbidity and Mortality Weekly Report 130
mosquitoes 6, 7, 22, 60, *60*, 62, 112–7, *113*
   nets 115, *116*
Mount Vernon 54
Muller, Paul 115
mumps 46
Munich sewer system *103*
mutations 20, 107, 108
mycobacteria 70, 73
Mycobacterium tuberculosis 70, 100

N

national boundaries 155

Native Americans 46
natural disasters 129
natural selection 18, 20, 22
Neisser, Albert 121
Neisseria gonorrhoeae 121
Nelmes, Sarah 65
Nightingale, Florence 90–93, *90, 92*
   Jewel *90*
   Medal 92
   Pledge 92
   School 90
Nobel Prize 6, 7
Notes on Nursing 90
nutrition 90, 104

O

obligate intracellular bacterium 121
obstetrics 80
"Ode on a Grecian Urn" 77
"Ode to a Nightingale" 77
Osler, Sir William 76, *76*

P

Paine, Thomas 58
Panama Canal builders 60, *63*
Pap test 140
Papanikolaou, Georgios 140, *140*
papillomas *see* skin warts
papillomavirus 13
parainfluenza virus 8
parasites 8, 10
Pasteur, Louis 86, 94–7, *94*, 98, 101
Pasteur, Marie 94
   Institute 94, 96, 130, *134*
Pasteurization 94
Pathogens 15, 100, 142, 144
PCR 13
Peking University 151
Peloponnesian War 28, *29*
penicillin 19, *19*, 118, 122–5, 126
   mould 122, *122*
penicillium 122
Pennsylvania Hospital 58
Pennsylvania, University of 58
peptic ulcer disease 136–9, *136*

Index

Pericles 29, 31
pertussis 8
Pettenkofer, Max 102–3, 103, 136
phagocytosis 32
phenol 86, 89
Philadelphia 58, 59, 60
phlebotomy 62, 63
phlegm 25, 27
phthisis see tuberculosis
physicians 24, 25
plague 28, 37, 37
    of Athens 28–31, 29, 30, 31, 38
    of Florence 38–41, 38, 41
plasmodium organism 113, 113, 114, 116
pneumonia 8, 9
    bacterial 104
    Pneumocystis 130, 132
polio 7, 126, 126, 128
    immunization 129
    vaccine 140
    vaccines 128
polymerase chain reaction see PCR
population spread 18–9, 152
postulates 100, 101, 136
Pott's disease of the spine 70
Poussin, Nicolas 31
probenecid 124
prokaryotes 10
prostration see sepsis
Protozoa 20
Protozoon 113
puerperal fever mortality rate 85
Pumphandle Lecture 83
pustules 64, 67, 68, 97
pyrazinamide 73

Q
quarantine 56
Quarantine Hospital, New York 57
Quarantine Question, The 55
quinine 116

R
rabies 94, 96, 97
ranitidine 139
Red Cross 92
nurse 106
Reed, Walter 7
resistance to antibiotics 19, 22
respiratory syncytial virus 8
respiratory viruses 154
retrovirus 132
reverse transcriptase enzyme 134
rifampin 73
RNA virus 60, 126, 132, 133, 146
Rome 78
Roosevelt, President Franklin 7, 128
Ross, Robert 6
Royal College of Physicians 80
Royal Society 51, 53, 67, 68, 89
Rush Medical College 60
Rush, Dr Benjamin 58–63, 59, 63

S
Sabin, Albert 128, 129
St Thomas' Hospital 90
Salk, Jonas 128, 140
salmonella 46
Salvarsan 120
sanitation 37, 70, 90, 102, 104
Santayana, George 152
sarin gas attack 145
SARS 10, 146, 146, 148
SARS-CoV 146, 148
SARS-CoV2 146, 146, 148
scrofula 100
self-medication 22
Semmelweis, Ignaz 84–5, 84, 85, 89
sepsis 85
Severe Acute Respiratory Syndrome see SARS
Severn, Joseph 78
sexually transmitted infections 118–21, 141
Shanghai 151
Shelley, Percy Bysshe 76, 76, 78
shingles 16
sickle cell anaemia 20, 22
sickle cell distribution 23
Siderophore 16
Silent Spring 115
silkworms 94, 96, 97
silver nitrate 121
single-celled organisms 51
skin warts 141
slums 152
smallpox 6, 7, 37, 45, 46, 54, 54, 55–6, 64–9, 65, 126, 142, 144
    biological weapon 46
    eradication 46
    mortality rate 65
    spread 48–9
    virus 143, 154
Snow, John 6, 80–3, 80, 83, 101, 126
social distancing 151, 151
Soho 80
South Sudan 154
Spain 104
Spanish flu 104–11, 104
    prevention chart 108
    second wave chart 110
Sparta 28, 31
speckling see variolation
spirochete 118
spontaneous generation 94
spread of plague 38–9, 40
spreading disease 152
sputum 73
staphylococci 122
Strasbourg 36
streptococci 84, 84
syphilis 10, 10, 118–20, 118, 126

T
tapeworm parasites 10
Taubenberger, Jeffrey 107
Tenochtitlan 44, 45, 47
tetanus 8
thalassemia 22
Theiler, Max 7
Thucydides 28, 29, 29, 30–1, 38, 40, 41
Tokyo subway 145
tranquilizer chair 63
Treponema pallidum 118, 118
Tu Youyou 116
tubercle bacillus 100
tuberculosis 54, 54, 70–5, 70, 73, 76–79, 98, 100, 100
    distribution 74
    of the spine 9
Tuskegee syphilis study 120, 121
"Typhoid Mary" 6
typhoid 94
typhus 37

U
United States of America 54
University College London 86

V
vaccination 73, 96, 114–5
variola see smallpox
variolation 56, 64, 67, 68
Vibrio cholera 102
Victoria, Queen 80
Vienna General Hospital 84
Virginia 56
virus 10, 14, 15, 16

W
War of Independence 55, 58
Warren, Robin 136–9, 139
Washington, George 54–9, 54, 57
Washington, Lawrence 54, 54
Wren, Christopher 53
Wuhan Province 149, 151, 154

X
x-ray crystallography 124

Y
yaws 126
yellow bile 25, 27
yellow fever 7, 58–63, 58, 126
Yersin, Alexandre 32, 32
Yersinia pestis 32, 32, 35

Z
zur Hausen, Harald 141, 141

CREDITS

ABOVE: Iranian firefighters in protective gear disinfect the streets of the capital of Tehran during the COVID-19 pandemic.

The publishers would like to thank the following sources for their kind permission to reproduce the pictures in this book.

Key: t = top, b = bottom, l = left, r = right & c = centre

Alamy: 7t, 7tr, 8, 25, 26, 28t, 30, 32, 33, 36b, 38, 45, 51, 52, 53, 54b, 55, 56t, 56b, 57, 61, 62, 74, 75b, 76, 78, 79, 81t, 81b, 84t, 85, 95, 96, 103l, 103r, 105, 121tr, 124, 130, 131, 132, 133, 134, 135, 140c, 141l, 146, 148b

Bridgeman Images: 39, 40, 41, 97b

Photo by Dan Burton on Unsplash: 150t

Centers for Disease Control and Prevention: 9, 148t

Getty Images: 19c, 19b, 20, 27, 31, 34, 36t, 44, 46-47, 59t, 59b, 75b, 92, 97t, 99, 106l, 116, 117t, 117b, 119, 120, 121t, 121c, 122, 123, 125t, 125b, 127, 129tr, 129c, 131, 132, 133, 139c, 139b, 140b, 141b, 143tr, 144t, 144c, 145, 149, 150c, 151, 154t, 155, 160

Heart of England NHS Foundation Trust. Attribution 4.0 International (CC BY 4.0): 71b

Library of Congress: 73

National Library of Medicine: 114

National Museum of Health & Medicine: 106b, 107

NIAID: 4, 147b

Public Domain: 28b, 40, 50t, 54t, 81, 84b

Photo by Tedward Quinn on Unsplash: 147t

Science Museum, London. Attribution 4.0 International (CC BY 4.0): 53b, 68

Science Photo Library: 11tl, 11bl, 14, 15t, 15b, 16t, 16c, 17, 21, 86, 112, 154

Shutterstock: 11tr, 63, 75, 106t, 109r, 113, 116b, 118, 143tl, 144b, 152, 153

Wellcome Collection. Attribution 4.0 International (CC BY 4.0): 12, 19t, 24, 27b 35t, 35b, 42-43, 50b, 58, 64t, 64b, 66, 67, 69, 71tl, 71tr, 72, 76c, 87, 88, 89, 90t, 90b, 91, 93, 98, 100, 101, 128, 129tl

Every effort has been made to acknowledge correctly and contact the source and/or copyright holder of each picture and Welbeck Publishing apologises for any unintentional errors or omissions, which will be corrected in future editions of this book.